Also by Thomas Homer-Dixon

*The Ingenuity Gap* (2000)

*The Upside of Down* (2006)

# COMMANDING HOPE

## THE POWER WE HAVE TO RENEW A WORLD IN PERIL

THOMAS HOMER-DIXON

ALFRED A. KNOPF CANADA

PUBLISHED BY ALFRED A. KNOPF CANADA

 Published in 2020 by Alfred A. Knopf Canada, a division of Penguin Random House Canada Limited, Toronto. Distributed in Canada by Penguin Random House Canada Limited, Toronto.

www.penguinrandomhouse.ca



Pages 424 to 427 constitute a continuation of the copyright page.

Library and Archives Canada Cataloguing in Publication

Title: Commanding hope : the power we have to renew a world in peril / Thomas Homer-Dixon.
Names: Homer-Dixon, Thomas F., author.
Identifiers: Canadiana (print) 20200199579 | Canadiana (ebook) 20200199633 | ISBN 9780307363169 (hardcover) | ISBN 9780307363183 (EPUB)
Subjects: LCSH: Environmental responsibility. | LCSH: Creative ability. | LCSH: Social change.
Classification: LCC GE195.7 .H63 2020 | DDC 363.7—dc23

Text design: Lisa Jager
Jacket design: Lisa Jager
Image credits: (trees) © simonlong / Moment / Getty Images; (starry sky) Shotaro Hamasaki / Unsplash

Printed and bound in Canada

2 4 6 8 9 7 5 3 1

Penguin Random House
KNOPF CANADA

For Ben and Kate

*Nobis non desistendum est*

Hope is stubborn. It exists in us at the cellular level
and works up from there, as part of the urge to live.
So hope will persist. The question is, can we put it to use?

KIM STANLEY ROBINSON

## CONTENTS

# Prologue

# Our Own Story

*When we get our story wrong, we get*
*our future wrong.* David Korten

"WHAT'S THIS STORY ABOUT, MUMMY?"

Our four-year-old daughter, Kate, had found a printed copy of an article on my wife's desk. But she couldn't understand its bold title: "Approaching a State Shift in Earth's Biosphere."

It was early June 2012, and the article had just appeared on the website of the top scientific journal *Nature*.[1] News about it was flashing through networks of scientists, academics, and others interested in how our planet is changing. Its twenty-two authors, among them some of the world's leading ecologists, had reviewed findings on how forests, grasslands, and other natural landscapes react when severely disrupted.

They reported that when at least half of a landscape's area has undergone major change, its undisturbed parts are far more likely to flip suddenly to a totally new state. Logging more than half a forest while leaving behind only fragments of undisturbed forestland, for example, dramatically boosts the risk that the populations of plants and animals in those remaining fragments will crash. The scientists then used the latest data on our human species' growing numbers, economic activity, and global impacts like climate change to estimate when, on the current course, Earth's total landscape will cross that 50 percent threshold.

Their conclusion: somewhere around 2045.

My wife, Sarah, a professor of geography, took a deep breath before answering: "It's about how the world might change when you're a little bigger, darling."

"Okay."

Feeling sad, Sarah went back to grading papers. But Kate flipped the article over to the blank side of the back page. Helping herself to two colored pens, she drew a picture of a green-and-red flower growing out of a green landscape. She gave the flower a red happy face. Underneath, she drew a little stick figure of herself, also with a happy face, waving at the viewer.

Kate was telling her own story about what she wanted the world to be like when she was bigger.

### REACHING FOR LIGHT

Maybe you feel it too: a creeping sense that the world is going haywire. A darkness spreading across the horizon of our aspirations for our families, our communities, our world. An emerging dismay that possibilities for a good future, for ourselves and our kids, are ebbing away.

If so, your feelings are not without base; they do reflect a real shift in the state of our world. Accumulating scientific evidence and data show that key trendlines gauging humanity's well-being—economic, social, political, and environmental—have indeed turned sharply downwards.

Just twenty years ago a feeling of exuberance still animated many societies. After the Soviet Union collapsed and before the war on terror, political, business, and intellectual leaders in the West declared that a fusion of capitalism, liberal democracy, and modern science would create a future of near-boundless possibility for all humanity. Now, humanity is at a perilous juncture. Problems like climate change, economic and social inequality, and the risk of nuclear war have become critical. In 2020, COVID-19 stopped the world at large in its tracks. International scientific agencies are issuing report after report declaring that a global environmental catastrophe is imminent, now probably far earlier than 2045, and maybe even as soon as a decade from now. Meanwhile, reason and scientific fact often seem impotent before entrenched vested interests, worsening social polarization, and rising political authoritarianism.

As our prospects seem to diminish by the day, some of us retreat inwards to focus on things close to us in time and space, such as our friends and family, in person and on social media. Others try denial, maybe by claiming that the evidence for problems such as climate change and even pandemics is invented by people who benefit from scaring us. Or, we become fatalistic, declaring we can't do anything about the problems because we've gotten used to a way of living or because the problems are the fault of the rich, or the poor, or immigrants, minorities, or "them over there"—anybody but us. Some of us rally to authoritarian leaders who tell a simple story about what's wrong and declare they can make things better with bold, harsh action.

Anxiety about the future, detachment, self-deception, and feelings of resentment and helplessness—this is a perilous psychological state—the starting line of a fast track to the end of hope. It also makes the future we fear far more likely to happen, because the best way to ensure we'll fail to solve our problems is to believe we can't.

We all know—whether explicitly or unconsciously—that to escape this trap we need to come up with promising ideas to address the critical problems humanity faces. But to do so, we need to understand what's causing the problems in the first place. As any medical doctor would say, good prescription depends on good diagnosis. To that end, over the last forty years I've studied humankind's global challenges closely, particularly worsening economic insecurity, climate change, pandemics, scarcities of critical resources like fresh water and clean energy, weak and incompetent governance, and the factors that keep our societies from innovating effectively to address such problems. I've also studied how these challenges can combine to multiply their total impact, with cascading consequences that sometimes lead to mass violence, including terrorism, genocide, and war.

As a doctoral student at the Massachusetts Institute of Technology in the 1980s, I helped found a research group of young natural scientists, lawyers, and social scientists at MIT, Harvard, and other nearby universities interested in the implications of Earth's environmental crisis. Our work together was exhilarating—we were all hopeful that science, international goodwill, and basic common sense would prevent humanity from tumbling into an environmental disaster. Today, we're dispersed all over the world; and with a planetary environmental disaster now unfolding in real time, we remain connected with each other to share information, ideas, and research findings.

Alas, the underlying causes of humanity's problems aren't easy to diagnose, and some of the world's best minds have struggled for decades to figure out what's going on. Ever since those university years, I've followed their research and expert debates with fascination, and my books *The Ingenuity Gap* (2000) and *The Upside of Down* (2006) drew on that work to provide a framework for my own research and diagnoses.[2] I didn't pull any punches in my assessment of the dangers, so I was often labeled a "doom-meister." But as the years have passed, my analysis in those books has (unfortunately) turned out to be close to the mark, and the profound gravity of humanity's predicament is now hard to miss and broadly acknowledged.

I've always intended this third book to move beyond diagnosis to explore what we can do to get through the gloom and reach a new light. I start from the assumption that this is a time for honesty about the challenges we face and about our need for immediate, courageous responses. It's now vividly apparent to me and my scientific colleagues, to many members of the world's Indigenous cultures, to socially progressive groups everywhere, to the clear-eyed youth who in 2019 protested for climate action in the streets of more than a hundred countries, and to the countless families and communities worldwide devastated by the psychological and economic trauma of the COVID-19 pandemic the following year, that humanity is marching down a path towards calamity. To find a route to a far better outcome, we must marshal our amazing ability to overcome new challenges—an ability we've honed since the first hominid climbed down from the trees and set out across the savannah.

In the following chapters, I draw on insights from history, psychology, physics, philosophy, economics, politics, and art to identify such an alternative route that's informed by honest realism—one that leads us towards a future of broadly shared opportunity, security, justice, and identity. I also provide some practical scientific tools that we can use to take our first steps together along this radically new path.

I argue that at this crucial moment in humanity's history, three changes are essential to keep us from descending into intractable, savage violence.

First, we need individually to better understand how and why we see the world the way we do and what makes other people's views sometimes so different from ours. Second, instead of passively accepting a dystopian image of what will come tomorrow, we need to actively create together from our diverse perspectives a shared story of a positive future—including a shared identity as "we"—that will help us address our common problems and thrive.[3] And, finally, we need to fully mobilize our extraordinary human agency to produce that future.

Each of these changes requires that we have hope. To believe in the possible and to make the possible real, we must recognize that the right kind of hope can be a tool of change, and we must give our hope the muscle it requires in our present crisis.

Unfortunately, though, hope has seen better days. Barely more than a decade ago, Barack Obama could speak unabashedly of the "audacity of hope" in his presidential campaign, and his idea was a powerfully motivating psychological and social force in the world. And over the last fifteen years, eminent thinkers and social scientists have called for "radical hope," "active hope," and "intrinsic hope."[4] But despite these vital efforts to rejuvenate the idea, many of us have come to regard hope with disdain—as a state of mind that's naive and irresolute at best, delusional at worst.

Yet if we're to survive, let alone see our children prosper in this century and beyond, we need a potently motivating principle that's honest about the gravity of the dangers we face and about the personal responsibility each and every one of us has to face those dangers; that's astute about the strategies we can use to overcome those dangers, given the viewpoints, values, and goals of people around us; and that's powerful because it galvanizes our agency, our capacity to discern our most promising paths forward and choose among them. We need, in other words, the kind of hope that has motivated millions of young climate activists to sit outside parliament houses and block business-as-usual traffic in capital cities worldwide and that has galvanized communities and nations around Earth to slow the coronavirus pandemic.

In Dante's fourteenth-century epic poem *The Divine Comedy*, the entrance to Hell famously carries the inscription: "Abandon all hope, ye who enter here." The phrase has become watered down over time, almost trite. But facing a future that promises to be hell for countless people, our task in the twenty-first century is to rediscover the power of the uniquely human ability to hope—an ability to envision and strive towards a positive future that's an alternative to whatever challenging or even unbearable present we're living in.

I propose in the pages that follow a way of mobilizing hope's immense psychological power, as people have done in times of great stress before and can do again. What I call *commanding hope* is grounded in historical and scientific knowledge of how hope works at every level—in our lives as individual human beings and in our societies too. Today, confronting challenges so large that all too often we feel unable to move, we need it more than ever.

There are no guarantees of success. The perils are real, and the chances we'll prevail may be small. But we face a choice between denying reality, running from the crisis, or facing that crisis head on to fight for a far better future. I've written this book for all of us—community activists, parents and grandparents, students and teachers, business and religious leaders, farmers and builders, scientists and engineers, nurses and doctors , restaurant and shop owners and artists, politicians and voters—*all of us* who choose to fight.

And it's dedicated to my children, Ben and Kate, and through them to all the children who remind us every day how to use our imaginations to tell our own story, and to see and seek the world we want.

# PART ONE

# THE NECESSITY OF HOPE

I

# Signals

*There are dark shadows on the earth, but its lights are stronger in the contrast.* Charles Dickens

**SPRING 1957, BLOOMFIELD, CONNECTICUT**

A little girl, three years old, plays with a toy rotary phone on the kitchen floor. "I want to speak to ministers, priests, and rabbis," she says into the mouthpiece. "I have a 'tition."

Her mother, Stephanie May, sits at a Formica table nearby. Dialing the wall phone beside her, she calls one prominent local religious leader after another. "Hello, I'm a volunteer with the Connecticut Committee to Halt Nuclear Testing, and I wonder if you'd be willing to circulate our petition?" After each call, she pencils a note on a list of names.

Outside the farmhouse's kitchen window, a rough lawn slopes away to some woods. The sun's morning rays catch wildflowers scattered in the lawn and evaporate the dew into a faint mist, while an occasional barn swallow darts among a cluster of paper birches and a twisted old willow. Nearby sits a small playhouse, built by the girl's father—a magical place constructed with live-edge planks and sheltering some chairs and a little table laid with a miniature tea set.

But this blissful scene couldn't be further from Stephanie's thoughts as she makes her calls. In her mind are images of huge fireballs obliterating faraway landscapes and the dread of childhood leukemia that could afflict her own home. She's determined to protect her daughter and baby son. So she plans to throw a wrench into the machines of war.

These machines, especially the militaries of the United States and Soviet Union, commanded almost incomprehensible destructive might. They'd emerged from a world shattered by a string of catastrophes spanning four decades that had pummeled people's spirits, including economic depression, two world wars, the gas chambers of Auschwitz, and totalitarian efforts to dominate the planet. By 1950, many people awaited a third world war, this time nuclear, between East and West—and felt gnawing uncertainty, fear, and, too often, barely repressed despair. The great political philosopher Hannah Arendt captured the moment's sentiment brilliantly, with language that resonates eerily today. "Never has our future been more unpredictable," she declared. "Never have we depended so much on political forces that cannot be trusted to follow the rules of common sense and self-interest—forces that look like sheer insanity, if judged by the standards of other centuries."[1]

Just two years after Arendt wrote those words, the United States detonated its first hydrogen bomb in the equatorial Pacific. The "Ivy Mike" test was seven hundred times more powerful than the Hiroshima bomb and generated a cloud forty kilometers high and 160 kilometers wide. Fourteen months passed, and then another US hydrogen bomb, code-named "Castle Bravo"—this one

*Ivy Mike, November 1952*

50 percent more potent than Mike—blasted apart a coral island in the Bikini Atoll. "The entire island was extirpated from the sea, leaving a gaping hole in the ocean floor," two scientists reported. "The evaporated metal of the test structure, the pulverized concrete underneath, and millions of tons of sand, coral, and sea water were sucked upward by the ball of fire as it roared into the sky."[2] Soon afterwards, the Soviet Union answered with its own immense hydrogen bomb at Semipalatinsk.

Although by 1957 the popular mood in the United States was more upbeat than it had been in many years—the country was finally at peace and its economy surging—in the background, Arendt's insane political forces remained hard at work. The United States, the Soviet Union, and the United Kingdom had by then detonated a total of 125 nuclear bombs in the atmosphere. Each explosion, experts later explained "resulted in unrestrained release into the environment of substantial quantities of radioactive materials, which were . . . deposited everywhere on the Earth's surface."[3]

Accomplished sculptor and pianist, member of the Junior League, and just thirty years old, "Mrs. John W. May" was patronized as a "Bloomfield housewife" by the local *Bridgeport Post.* Still, she'd learned from newspaper reports that the tests were covering the landscape around her community with tiny radioactive particles. The particles settled on the grass and fodder eaten by local cows, and the cows' digestive systems concentrated the radioactivity in their milk, which children then drank. The most dangerous component was strontium-90, an isotope that mimics calcium in the human body and accumulates in children's bones and marrow as they grow. Its radioactivity sometimes causes leukemia and bone cancer.

Stephanie was appalled, but she wasn't going to let fear paralyze her. She started sending letters to the local newspaper, and before long another housewife, Virginia Davis, wrote to ask if she could help. Then Stephanie and her husband, John, along with Virginia and her husband, Martin, created a committee—the Connecticut Committee to Halt Nuclear Testing. Stephanie and Virginia were co-chairs, and John was treasurer and secretary. They paid a local firm to print a petition; its preamble laid out the basic scientific facts about the dangers of nuclear fallout and appealed, as Stephanie's daughter wrote later, "to a moral objection to poisoning the land and killing innocents without the permission of anyone, least of all the innocents."[4]

In those days, getting lots of people to sign a petition wasn't just a matter of a few keystrokes and clicks on a website. Stephanie and Virginia opened the yellow pages of their phone books at *Clergy* and, starting at the letter *A*, called every minister, priest, and rabbi in the vicinity, asking them to distribute the petition to their congregations. In a little over two weeks, after making over three hundred calls, they'd gathered two thousand signatures from regular citizens and prominent community members, including scientists, doctors, and professors.

In the next months, their little committee grew astonishingly quickly, linking with other groups started by mothers and activists across the country and overseas. Soon, Stephanie was asked to join seventeen plaintiffs—including two Nobel Prize winners—in an anti-testing lawsuit against the governments of the United States, Soviet Union, and United Kingdom; and she was invited to sit on the board of the National Committee for a Sane Nuclear Policy, along with some of the country's most prominent authors, publishers, and intellectuals. Meanwhile, hate mail filled her farmhouse mailbox, and threatening phone calls arrived late at night. "Your house is surrounded. Come out with your hands up."

Despite fierce opposition from some members of the public—and from many powerful officials, experts, and politicians too—the movement snowballed, as more and more ordinary people got involved. And governments began to pay attention. In 1958, the three nuclear-armed powers started negotiating a test ban and agreed to a temporary testing moratorium. But the protests continued, because people sensed they had to keep pressure on governments. On Easter weekend of 1960 in London, England, at the end of a march from the British Atomic Weapons Research Establishment at Aldermaston, Stephanie addressed a huge crowd. Speaking from the base of Lord Nelson's column in Trafalgar Square before one hundred thousand people, her six-year-old daughter Elizabeth

beside her, she announced that she brought support from mothers and concerned citizens in the United States. It was, according to the BBC at the time, the largest demonstration London had seen in the twentieth century.

The test-ban negotiations among the three powers were fitful, repeatedly running aground on technical complexities and often derailed by international crises. In late 1961, the moratorium collapsed amidst bitter recriminations between the United States and Soviet Union. A monstrous binge of testing ensued, involving a stunning 179 atmospheric blasts in just two years. But in October 1962, during the Cuban Missile Crisis, the world looked directly into the nuclear abyss, and in the summer of 1963, the United States, Soviet Union, and United Kingdom finally signed the Partial Test Ban Treaty—moving all nuclear testing underground.

The treaty wasn't perfect. Nuclear weapons research and development went on unabated; and France and later China, to their enduring discredit, kept testing in the atmosphere well into the 1970s. But it was the first major arms-control agreement between the nuclear powers and among the first global environmental agreements.[5] Concentrations of radioactive material in the air began declining immediately. Some key isotopes returned to near-normal levels in twenty years.

Stephanie May, her fellow mothers, fathers, children, activists, and protesters in many countries, along with concerned scientists, test-ban negotiators, and key visionary political leaders had accomplished something truly remarkable. Their accomplishment was a product of their hope, just as it was an enormous source of hope for others too. But the full story of that hope—its origins, its true meaning, the source of its power, and why we urgently need something similar today—must still be told.

**JULY 5, 2015, VANCOUVER ISLAND**

The change came in barely fifteen minutes. I was standing on a cliff overlooking the Strait of Juan de Fuca that separates the extreme western part of Canada from the United States. Before me was a panoramic view of Washington State's Olympic Mountains rising out of the sea twenty kilometers away. Then, I glanced straight up.

A dense cloud of smoke had appeared abruptly in the sky overhead. The cloud's smooth edge, highlighted by the afternoon sun, was sweeping the heavens like a gigantic scythe. Behind that edge, the sky's brilliant blue had turned a murky brown. Soon, the cloud transformed the sun into a feeble orange disc, dusk fell in midafternoon, and the vivid colors of the sea and forest around me faded away.

**NOVEMBER 8, 2016, SOUTHERN ONTARIO**

Many of us retain a vivid picture in our minds of the very moment we realized something momentous was happening. Psychologists call such mental images "flashbulb memories"; they're sharp recollections of the instant we heard a startling piece of news or understood an astonishing and incontrovertible fact.

Along with hundreds of millions of people around the world, I'd been watching the US election returns. Early results were generally in line with predictions. But by 9:15 p.m., Donald Trump was polling far more strongly than had been expected. The liberal news commentators were still saying that Hillary Clinton was likely to prevail, but in that flashbulb moment I felt certain of the result.

Trump's win didn't really surprise me though, because for years I'd tracked how social, economic, and political stresses were building in the United States. My reading of the best research on the causes of political extremism, social instability, and civil violence told me that the country was primed for the rise of an authoritarian leader. I reject the values, beliefs, and policies Trump represents: his rabble-rousing demagoguery, attacks on journalists, and open contempt for the rule of law and checks on executive power mirror dangerous patterns that anyone familiar with modern Western history should recognize. But on the night of his election, I was perplexed by what those of us who are opposed to this kind of social and political program, both inside and outside the United States, should do next. Trumpian politics would propagate through societies, I believed, pushing the world into a spiral of ever-worsening rancor between mutually uncomprehending groups—a spiral that could ultimately drive us all towards violence and even the end of freedom, democracy, and the equalizing dream of prosperity for everyone. How, I asked myself, could humanity create enough commonality of goals, beliefs, and values to act positively together in a way that benefits all?

### AUGUST 2, 2017, VANCOUVER ISLAND

This time the change came more slowly than it had two years before, in July 2015, but the results were identical. Standing on the same cliff, I could see smoke creating a faint horizontal line above the mountains across the Strait, tracing a previously unseen boundary between atmospheric layers. Gradually the line thickened into a cloud, and then the cloud expanded into an immense filthy mass that extended from the mountains towards me, over my head, and to the horizon beyond my back. For a while, a bright band of sky remained visible over the mountains, with the cloud's blackness floating above. But then the mass pressed downwards, snuffing out the final band of open sky and descending relentlessly from the mountains' peaks down their forested slopes. I could see and feel the smoke pressing down on me too. When it finally reached sea level, the cloud's acrid haze enveloped me. The air stank of soot. The mountains had vanished entirely, and I couldn't differentiate between ocean and air. The buoys of crab traps only two hundred meters offshore—minutes before clearly distinct—were now faint specks.

The smoke told a story: distant forests were burning. Hundreds of kilometers away, vast tracts of pine and fir forest in British Columbia's central regions—scorched by weeks of record midsummer temperatures, and parched by fast winds and ultra-low humidity—had exploded into flames, displacing tens of thousands of people and destroying hundreds of homes, ranches, businesses, and industries. At first, onshore winds from the Pacific had pushed the smoke into the continent's interior. But now winds were blowing in the opposite direction. Huge airborne rivers of smoke and ash were flooding out of central BC, following the land's rugged contours and its great river valleys down to the sea, smothering the coastal metropolis of Vancouver, spreading into Puget Sound and southwards to Seattle, and spilling across Vancouver Island and into its adjacent straits. Within just two days, satellite images showed nearly a thousand kilometers of North America's Pacific coastline obscured by smoke.

*Southern British Columbia coastline and Puget Sound from space, August 4, 2017 (NASA)*

The cloud stayed anchored in place for weeks—still, silent, and omnipresent. The smoke-drenched landscape looked like a scene from a dystopian fantasy. Color and even oxygen seemed to have been sucked from daily life. Buildings, streets, and landscapes were shrouded in an ochre gauze, indistinct and dreamlike. Sometimes ash rained from the sky. Where once the air had been an invisible and taken-for-granted source of life, now it was constantly perceptible and menacing; hospital emergency rooms filled day after day with people in respiratory distress.

Twice in two years was twice too much. Old-timers with memories going back nearly a century had never seen a single episode of the kind, let alone two, and the second was far harder to dismiss as a bizarre anomaly than the first.[6] Then, as British Columbia's forests continued to burn, reports arrived of horrific heat waves, droughts, and wildfires across the western United States, and in Portugal, southern France, and Croatia. An unspoken question hovered in many people's minds: "Is this what our future will look like?"

Less than three years later, an altogether different kind of event changed nearly everyone's image of the future.

### FEBRUARY 22, 2020, DATELINE HONG KONG

"Surge in Cases Raises Concern of a Pandemic" —*New York Times*.

> An alarming surge of new coronavirus cases outside China, with fears of a major outbreak in Iran, is threatening to transform the contagion into a global pandemic, as countries around the Middle East scrambled to close their borders and continents so far largely spared reported big upticks in the illness.
>
> In Iran, which had insisted as recently as Tuesday that it had no cases, the virus may now have reached most major

> cities. . . . Already, cases of travelers from Iran testing positive for the virus have turned up in Canada and Lebanon. . . .
>
> The United States now has 34 cases, with more expected, and Italy experienced a spike from three cases to 17 and ordered mandatory quarantine measures.[7]

### BENDING THE CURVE

Nuclear weapons' dangers. Donald Trump's 2016 election. Vast clouds of wildfire smoke. A global pandemic.

These four things seem entirely different, but they share one key similarity: each signals that something is going awry in the story of human progress.

Of course, that story is enormously complex and can be read in many ways. By one reading, our species has made stunning advances since Stephanie May dialed her kitchen phone more than sixty years ago. As authors such as Steven Pinker note, rates of wretched poverty, infectious disease, and mass violence have all plummeted around the world in recent decades.[8] And when we look ahead, exciting technological advances in genetic medicine, robotics, artificial intelligence, and the like seem about to bring huge improvements to (at least some) people's lives.

But by another reading, humanity is no longer moving towards a future anymore that will allow our communities locally, nationally, and globally to flourish together. If we're moving anywhere at all, we seem to be sliding down a slippery slope towards a treacherous place, with unreason, self-righteous anger, and racism greasing our shared descent.

Nations have abjectly failed to rid the world of nuclear weapons once and for all. The great powers (and a few smaller ones too) still deploy these hideous machines by the thousands, many on high alert, despite countless mass protests down through the years, as well as advocacy campaigns, desperate pleas by scientists, and earnest commitments by statesmen. Worse, after several decades when treaties and détente between nuclear-armed countries did genuinely reduce the risk of nuclear war, it's now escalating again, as the United States pulls out of arms-control treaties, while leaders of the United States, Russia, North Korea, India, and Pakistan brag about using nukes to annihilate their enemies.

Across other fronts, too, threats to our collective well-being are multiplying. We see economic, political, and social distress soaring within many societies from widening inequalities of wealth and opportunity, large disparities in human population growth between the world's rich and poor regions, surging flows of distressed migrants and refugees inside countries and across international borders, and cascading effects of contagious disease—in 2020, hundreds of millions of jobs lost, countless companies bankrupt, major countries defaulting on their debts, and financial crises rolling back and forth through world markets. We're seeing, as well, worsening instabilities in key natural systems, particularly Earth's climate. As humanity's staggering resource consumption and pollution output push these natural systems out of equilibrium, coral reefs die; populations of vertebrates, birds, and insects, including pollinators, plummet; the Arctic ice cap shrinks; extraordinarily powerful storms ravage the world's coastlines; droughts bake its croplands; and vicious fires rip through its forests.

We might consciously notice some of these threats—as when news of cross-border migrations or economic dislocations directly affects us, smoke blots out the sky overhead, or a new coronavirus locks down our community, costs us our jobs and destroys our savings—but most of us are only subliminally aware of the other threats, like economic inequality and disappearing pollinators. Still, even this subliminal awareness can powerfully affect our general mood by creating an intuition that things are going wrong—that our lives are ever more insecure, that other people and groups are getting ahead of us, and, perhaps most alarmingly, that our children are likely to be worse off than us when they grow up. That intuition is reinforced by an apparently nonstop stream of events—headlined in our news feeds—that not long ago would have seemed, once again, right out of a dystopian fantasy. We see video clips of young men driving trucks at high speed into families strolling in city streets; an accountant firing automatic weapons from a hotel tower into a crowd at a music festival; professed Nazis marching shamelessly in public spaces; and war combatants in Syria deliberately targeting hospitals with missiles and children—children!—with poison gas.

As the feeling that things are going wrong becomes palpable, people become more likely to perceive the world as fraught with uncertainty and bounded by constraint, and they're more likely to think that danger and awful surprise lurk just beyond the edge of their day-to-day reality. And as they become hesitant and anxious—as their faith in progress shrinks—they become more inclined to talk to each other in the emotional cadences of doubt and fear, and the stories they tell about their future become muted and drained of excitement and positive possibility. (The French word *inquiétude*—a disturbing emotional state caused by fear, worry, and uncertainty—captures this feeling well.)

Mood matters. When hundreds of millions if not billions of people start to feel these ways, the basic dynamic of humanity's politics can shift abruptly—as we saw, for instance, during the Great Depression and its aftermath.[9] Many people, scared and resentful, turn away from leaders who seem soft, incompetent, and beholden to powerful elite interests, and unable to come to grips with the problems affecting them, and turn towards those who are hard, angry, and decisive (yet often equally or more beholden to elite interests) and who declare they'll protect families, communities, and nations with whatever means necessary.[10] Authoritarianism gains ground.

Autocrats have historically played on their followers' social and economic insecurities and on fears of "them" to curtail basic political and civil rights, suffocate free speech and assembly, and jail or sometimes assassinate opponents and journalists. Today, even in some of the world's oldest liberal democracies like the United States and England, populist politicians and their enablers are stoking grievances towards outsiders; attacking the free press, basic democratic norms of tolerance and forbearance, and the rule of law; and denying well-established scientific fact.

This shift is fraying the fabric of norms, institutions, and laws—the liberal international order—that people worldwide have laboriously woven since the middle of the nineteenth century in innovative answer to world wars, genocides, financial crises, famines, pandemics, and environmental calamities. It ranges from the Geneva Protocol banning the use of poison gas to the UN Convention on the Rights of the Child, the World Health Organization, and the Paris Agreement on climate change. Threadbare even at its best, this fabric is now being shredded by neglect, underfunding, the outright hostility of many nations and, most importantly, the impunity enjoyed by those who regularly violate its makeshift moral framework.

What does all of this portend? Are we at a point where the curve of our collective well-being begins to bend sharply downwards? It appears so: when we dispassionately weigh the evidence, the negative trends I've just described seem to be starting to overwhelm the positive ones highlighted by Steven Pinker, Gregg Easterbrook, Matt Ridley, and like-minded commentators. This new downward trend isn't inexorable, though . . . at least not yet. There are still many things we can do to bend our curve upwards again.

And by "we" here, I mean not just North Americans or members of the richer Western societies, but all people, everywhere. The entire human species is involved today; our fates are too intertwined on Earth for it to be otherwise, as the COVID-19 pandemic's explosive propagation through our tightly connected global networks has made abundantly clear. Humanity can't and won't address its urgent challenges unless enough of us from a broad range of cultures and societies recognize ourselves as one group, with a shared sense of identity, facing these challenges and developing solutions together.

## 2

# How About . . .

*Listen to the MUSTN'TS, child,*
*Listen to the DON'TS*
*Listen to the SHOULDN'TS*
*The IMPOSSIBLES, the WON'TS*
*Listen to the NEVER HAVES*
*Then listen close to me—*
*Anything can happen, child,*
*ANYTHING can be.*

Shel Silverstein

OUR DAUGHTER, KATE, is now twelve, and her brother, Ben, is fifteen. For as long as we could, Sarah and I tried hard to keep them in the magical bubble of childhood. Since we're both teachers and research academics, a prodigious quantity of information about current affairs and the state of the world flows through our household each day. This made it necessary to censor what Ben and Kate could see and hear: we limited screen time, kept smartphones and social media largely unavailable, and listened to newscasts mainly when the kids weren't around. When awful things happened in the world, which sometimes seemed to be daily, we turned newspapers over and tablets off.

But the bubble had to burst. As Ben and Kate got older, we knew we weren't doing them any favors by hiding them from the world or the world from them. As their social circles outside our home expanded, they inevitably heard more about what was happening in the world from friends, acquaintances, and teachers—some of it accurate, some of it sensational and wrong. They needed to learn more about what was happening to be prepared for what may be coming.

Yet how does one do that? I'm not even sure how to prepare myself.

What story should we tell young people who will become adults during what's likely to be an age of loss? What vision of the future should we create for them? How, specifically, does one tell a young person that the century during which they'll live could be marked by escalating insecurity, as a ferocious confluence of stresses—economic, demographic, environmental, technological, and political—tears at our societies, institutions, and basic civility and perhaps turns us viciously against each other? Even more, how can one help a young person, or anyone for that matter, psychologically cope with such knowledge so that they can do more than merely survive passively, but can flourish through a healthy, fulfilling, and humane life and help others to do so too?

These questions haunt me.

Children live through stories. They're always telling stories to themselves and other people, and they always want to know what stories the people and things around them are telling too. As Ben and Kate were growing up, they spent countless hours playing together. Like most children, they used their minds and toys to create stories about complex imaginary worlds. The stories unfolded in a stream of back-and-forth chatter. Every so often, one of them would introduce a new chapter with the words "How about. . . ." Sometimes Sarah and I would hear a "how about" every thirty seconds or so.

"How about the sea monster swims over to Atlantis and eats the submarine," announced Ben one day years ago. He was sitting with Kate amidst a jumble of blocks and bits of Lego representing the ruins of the lost submerged city. "Okay," said Kate, and they acted out the grim scene for a while. Then Kate decided to start her own chapter: "How about the sea monster gets a tummy ache and barfs the submarine into space!"

Ben and Kate's "How abouts" in these exchanges weren't questions—they were declarations. Each was a deliberate and decisive statement of possibility. Far more than adults, young children have a vivid sense of the constantly branching pathways of possibility as the instantaneous present moves inexorably into the future. And when a child chooses a pathway—signified by saying something like "How about" or "Let's pretend"—she starts a new chapter and turns possibility into reality in her mind.

Yet as children become adults, they give up most possibility and acknowledge that they're surrounded by limits. It's an essential and inevitable passage and, as all parents discover and many children come to recognize, a deeply bittersweet one too. On one hand, it's heartbreaking to see the flame of imagination and wonder fade as a young person leaves childhood behind. On the other, children must learn to distinguish quickly and accurately between what's possible and what's not. If they're to survive in a harsh world, they have to create a sharp line between imagination and reality and diminish the realm of imagination in their minds.

Still, in times of perplexing, unprecedented, and dangerous threats, we can learn much from childhood imagination—that wonderful capacity to generate endless "How abouts." For our species to succeed and prosper through this century, as it has during earlier eras of peril, we'll need to activate our ability to imagine possibilities.[1] Sometimes it's a mistake to fasten too quickly on constraints: it just makes those constraints appear inescapable and robs us of any chance for genuine change.

No matter how active our imaginations, however, we can't repeal reality. We still live in a physical or "material" world of rocks, trees, oceans, air, and light; and we can't wish away fundamental facts of this world, such as the laws of thermodynamics that describe how energy behaves. We also still live in a social world of loves, hatreds, group identities, business relationships, institutional rules, and formidable structures of money, power, and class. And, today, our indisputable reality again includes nuclear weapons placed on high alert, the ravages of a pandemic, extreme economic inequalities, worsening climate change, collapsing biodiversity, many forms of social and racial injustice, and rising authoritarianism.

So I live with one vivid fear for Ben and Kate—and for all children—as they emerge into adulthood and realize what this reality may mean for their future. I fear they'll lose hope, that their sense of future possibility—made vivid by their imaginations, and so wonderfully alive in their countless "How abouts" during their childhoods—will be crushed. Instead, they'll be led by the accumulating weight of evidence to tell themselves just one story about their future:

> We're all members of a failed species, and during this century we're destined to bear witness to the devastation of our planetary home and the violent unraveling of much of what we've accomplished.

If our children end up telling themselves only this story, at best they'll become fatalistic or morally pliant, or they'll escape into fantasy, or increasingly into depression, or live-for-the-moment partying. At worst, they'll yield to terror, rage, and brutal selfishness. They certainly won't work to find ways to transform their conditions to create a better world.

Of course, it's not just children who live through stories. Adults do too. They have their own stories about themselves, about the world around them, and about where they've come from and where they're going. When we ask what story we should tell young people about the future, or what vision of the future we should create for them, we're also really asking what story we should tell ourselves. In recent years, that question has regularly forced itself into my mind too—most often, somewhat strangely, in the middle of a January night.

### JANUARY RAINS

I tend to wake up in the early hours. I humor myself that this is as it should be: some brain scientists suggest that normal human sleep is bimodal, coming in two segments in the night. But during January nights over the last few years, my wakefulness has often been triggered by something bizarre: the sound of rain pounding on the roof overhead. The rain comes not on one night or two in the month, but sometimes night after night—pouring down in torrents, and often accompanied by thunderstorms that shake the walls and rattle the windows.

This kind of weather isn't remotely normal (not in January, at least) in the part of Canada where I live—the continent's Great Lakes zone, north of the forty-third parallel. Not long ago, rain rarely occurred in January. The month is the coldest of the year, and average temperatures should be well below freezing. It's a month when, on dark weekday evenings, kids in our town played hockey on backyard skating rinks under makeshift floodlights, and on weekends snowmobilers raced along winding trails through forests and farmlands. When we first moved to the area nearly twenty years ago, from late December into February the plows would create piles of snow two or three meters high along the sides of rural roads. In all of January, we might have had a day or two when it warmed a little above freezing—the "January thaw," as the locals called it.

I haven't seen big piles of snow beside the roads in years. With January temperatures sometimes pushing into the teens (in Celsius) and heavy rains lashing the countryside, whatever snow falls rarely accumulates. Most fathers have given up making backyard skating rinks, and snowmobile routes are often reduced to mud and bare sod.

In the middle of the night serotonin wanes and fretting waxes. As I lie in bed in January, the rain's sound fills me with dismay, because I know enough to recognize it as another signal of what's going awry in our world. I know that the amount of warming and climate disruption we're witnessing around the planet today closely tracks what climate scientists thirty years ago said would happen if we didn't rapidly curtail our fossil-fuel addiction.[2] Since they were largely right decades ago about what would happen now, we can objectively surmise they're largely right now about what's going to happen decades in the future.

And today's science is telling us that life will become phenomenally hard in a climate-changing world, if greenhouse gas emissions continue at anything like the current rate. Earth's average surface temperature has already risen above the maximum during the Holocene epoch—a geologic period going back to end of the last ice age, about twelve thousand years ago (see graph below). If carbon emissions aren't sharply cut, the world will see 2 to 3 degrees Celsius total warming before today's children reach old age. Such a change would vault Earth's average temperature far outside its range over the last two thousand years, when humanity laid down the infrastructure of modern civilization—its agricultural zones, water-distribution networks, roads, ports, major cities, and the like.

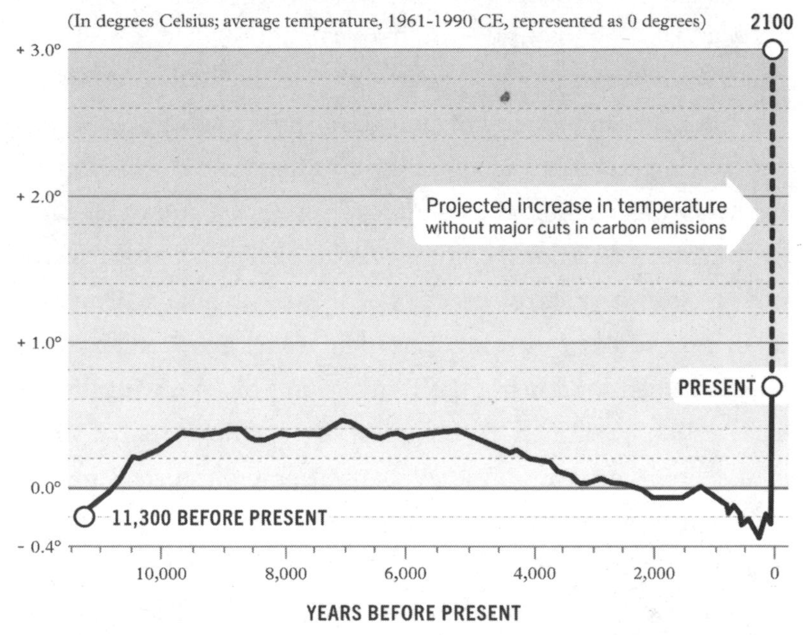


Some people find it hard to see why this is such a big deal. Sure, the average temperature of our atmosphere at Earth's surface is rising, but the incremental increase each decade—probably now around a quarter of a degree Celsius—seems tiny compared to the day-to-day and season-to-season temperature swings we experience all the time. In response to this skepticism, I've found I can better convey the peril we face by emphasizing less how we're warming the planet and more how we're altering *energy flows* around it.

As we boost the concentration of greenhouse gases in the atmosphere—mainly carbon dioxide from burning fossil fuels—we're making the atmosphere less transparent to infrared radiation (heat). So when the sunlight that falls on Earth's surface is converted to heat—you can feel this conversion when sunlight warms your skin—less of that heat escapes from the planet back into space. This extra heat is trapped in the atmosphere and oceans, and it amounts to about three-quarters of a watt per square meter of Earth's surface.

That quantity may seem miniscule, but it adds up to more than five thousand watts for each patch of Earth's surface the size of a football field. And when added up across the surface of the entire planet, it's equivalent to the energy that would be released by detonating about five hundred thousand Hiroshima-size atomic bombs in the atmosphere *every day*—each one releasing the explosive power contained in fifteen thousand tons of TNT.[3] That's one Hiroshima bomb's worth of extra trapped energy for each patch of Earth's surface the size of the US state of Rhode Island (the smallest state in the Union); about two bombs' worth for an area the size of the Canadian province of Prince Edward Island (the smallest province in the federation); sixteen for the Netherlands; and nearly one hundred for Britain . . . every day.

In this case, though, the extra energy doesn't pulverize and maim people, sicken them with radiation, flatten buildings, or start enormous fires like an atomic bomb does—at least not directly. Instead, it revs up the circulation of air and water around the planet, especially the hydrologic cycle, which is the cycle of water between Earth's surface and its atmosphere. It makes almost all our weather more extreme: storms are bigger, rainfall more intense, winds stronger, and droughts and heat waves more severe. It's as if we've created an enormous monster—a beast, like the mythical Leviathan or Behemoth—that's rampaging across Earth's surface. And as we pump more energy into this beast each year, it does only more damage.

Politicians and commentators often say that we're all simply going to have to adjust to the "new normal" of a climate-changed world. But there's no normal anymore, new or otherwise. We've already jumped from an equilibrium climate—the generally benign and stable climate that allowed our species to propagate and prosper over thousands of years, shown by the slowly undulating line across the bottom of the graph on page 32—to a climate regime that's always on the move, with temperatures shooting straight upwards. We're already courting the risk that this sudden change will unleash "runaway" feedbacks in Earth's climate, where warming's effects, like vast forest fires and melting Arctic permafrost, release huge quantities of greenhouse gases that then propel more warming. Once that happens, what humanity does to curb emissions won't matter much, because warming will have become its own cause.

The German climatologist and oceanographer Stefan Rahmstorf puts it bluntly: "We are catapulting ourselves way out of the Holocene." If humanity stays on its current climate trajectory, "we will not recognize our Earth by the end of this century."[4]

Hearing January's rain brings the science back to my mind. There's a stark difference, though, between understanding something scientifically and sensing it with one's own body and emotions. The former is an intellectual exercise, an exercise in solving a puzzle or satisfying one's curiosity. Hearing, and emotionally feeling, the rain turns the once-puzzling problem of climate change into a beast thrashing at the house's walls. If this is what's happening now, I wonder, what will it be like in a few decades, when Ben and Kate are my age? What are we doing to our world and its possibilities for our future?

### ANGER AND EXTREMISM

But at base I'm a scientist and not one to let either fretting or grief take over my life. As light dawns each morning in January, I have my own "Stephanie May" moment. I take my passive question—What story should I tell my children and myself about the future?—and turn it into an active question: What can I do to change humanity's course, even in the tiniest way, to make sure that future is brighter for us all? People everywhere who sense that the world is going haywire are asking the same question.

Writing this book has been part of my answer to that question. I started it three times—twice I wrote tens of thousands of words and then stalled. The two failed attempts taught me something unexpected and vitally important though: anyone who grasps the severity of humanity's predicament and tries to figure out how we might respond with something like a new organization, technology, or social movement to make things better—not just for ourselves narrowly, but for all of humanity—confronts an unforgiving conundrum, which I've come to call the *enough vs. feasible dilemma*. On one hand, changes that would be *enough* to make a real difference—that would genuinely reduce the danger humanity faces if they were implemented—don't appear to be *feasible*, in the sense that our societies aren't likely to implement them, because of existing political, economic, social, or technological roadblocks. On the other hand, changes that do currently appear feasible won't be enough by themselves.

In 2015, while I was struggling through my second attempt to write this book, my father died. He was a kind, wise, and enormously brave man—and a rock of stability in my life. During his last moments, I held him as his heartbeat faded away. I'm not a religious person, at least in the conventional sense, but the experience opened a window to something that seemed truly elemental and profoundly mysterious—as deaths of loved ones usually do. Oddly, though, it didn't strengthen a conviction that inescapable constraints, like death, govern our existence. Instead, it reinforced my intuition that we aren't remotely aware of much of reality's true substance. It seemed to create a filigree of hairline cracks in my image of reality, so that I sensed, behind the simulacrum of that image, a vast unknown world of unseen and unrecognized possibility.

Something else was on my mind that year too: a growing unease about the polarization and radicalization of the world's politics in country after country. In *The Upside of Down*, I'd challenged the assumption, common in the 1990s and 2000s, that capitalist liberal democracy was secure in already democratic countries and bound to spread to most of the world's non-democratic ones. I'd argued instead that conditions were ripening for an abrupt deviation towards extremist and authoritarian politics and weakening of democratic institutions, even in the West's long-standing democracies, as social stresses combined with feelings of powerlessness to create pervasive fear, frustration, and anger. But for such major transformation to happen, something other than just a shift in people's underlying emotional mood is needed, so I'd also proposed that the rise in political extremism would likely be preceded by severe social shocks, probably at the global level, and possibly first in the form of an international financial crisis. I'd highlighted the dangers accumulating from worsening financial imbalances between the United States and China and from speculation in the US housing market.

"People will want reassurance" in the wake of such a shock, I wrote. "They will want an explanation of the disorder that has engulfed them—an explanation that makes their world seem, once more, coherent and predictable, if not safe. Ruthless leaders can satisfy these desires and build their political power by prying open existing cleavages between ethnic and religious groups, classes, races, nations, or cultures."[5]

At the time, I expected that this abrupt move towards extremism wouldn't happen for a couple of decades or more—maybe sometime in the 2020s; I had the hunch it would take that long for stresses and shocks to erode our societies' natural coping mechanisms. But of course, a cruel shock did arrive only two years later, in the form of the 2008–09 financial crisis, which originated in the US housing market and quickly spread globally. The immediate emergency was stopped only when governments around the world undertook what was at the time the largest coordinated monetary and fiscal intervention in history.

Through the subsequent years of grinding economic disappointment in many countries, I waited for the other shoe to drop—the political radicalization shoe. Hints of a shift appeared sporadically in diverse countries—Thailand, Egypt, Turkey, Venezuela, India, Greece, Spain, Hungary, Poland, and then, with Brexit, the United Kingdom—but in each case, the causes were complex and dependent on context. And then in late 2016, Donald Trump was elected in the United States. Regardless of whether factors like Russian meddling had tipped the election's balance, here was just the kind of leader, mobilizing just the kind of popular anger, in just the ways that I and some fellow analysts had expected.

The election unleashed ferocious psychological forces. The rising discord, contention, and extremism that Trump and other populist leaders were injecting into world politics would now make implementing good solutions to the climate problem—or to other urgent problems our species faces—even less feasible. Scared, angry people aren't going to spontaneously sing "Kumbaya," hug, and then collaborate to fix such problems. We urgently need tools, I thought, to multiply our natural capacity to understand other people, allowing us to better see the world from their viewpoints, perhaps to find new ways of reaching rough agreement among ourselves or perhaps just to mobilize more effectively against the forces of reaction and division.

By a stroke of good fortune, I was already working with a remarkable group of researchers to develop tools of exactly this kind.

## THE GEOGRAPHY OF THE MIND

I'd long been fascinated by the links between psychological factors and violence among individuals, groups, and nations. But until the 2000s, my research had focused mainly on how changes in material factors—like the availability of fresh water, good soils, and energy resources—sometimes influence this violence, and on how societies innovate in response to such changes. I drew on ideas in the relatively new field of "complexity science," which aims to explain the inner workings of systems like forest ecologies, Earth's climate, the human brain, and human economies and societies. But since few universities have a critical mass of complexity researchers, I'd worked pretty much alone.

Then, in 2008, I moved from the University of Toronto to the University of Waterloo, where I discovered a research community bubbling with activity in this field. I was soon collaborating with two dozen scholars and doctoral students with expertise across more than ten disciplines. We quickly realized that we all shared an interest in applying complexity science to what goes on in people's minds, specifically to people's ideas about how the world works—their "worldviews." We found, too, that we were all worried about climate change and about the seemingly intractable polarization that was blocking good solutions to this problem. We decided to learn more about what role clashing worldviews play in this political mess.

In simplest terms, a worldview is a densely connected network of concepts, beliefs, and values in a person's mind. Each of us has one that's in some ways unique, since it's largely a product of our own firsthand experiences. A well-developed worldview gives us a mental framework to interpret things and events in our surroundings so that we can thrive. It gives us criteria for deciding what's good and what's bad, what's important and what isn't, and what actions will get us what we want and what won't. It also tells us where we are—most immediately in space and time, but also in our social milieu.

Our worldviews, then, provide the basic ideas from which we create our personal stories. In many ways, indeed, our worldviews create *who* we are—our personal identities, first as individuals, but also as members of groups ranging from our families to those associated with our gender, profession, and nation. Through our worldviews, we know what it means to be a member of a given group and how to distinguish people who belong from those who don't.

Working over several years, my colleagues and I invented two practical tools to help us understand what's going on deep inside people's worldviews.

The first was based on ideas that had been in the back of my mind for decades, but which finally clicked together when I saw them through a complexity lens. Called the "state-space method," it provides a way of identifying and analyzing key differences between worldviews. It builds on the concept of a state space, which is a picture, much like a three-dimensional map, of all possible states of a given system, such as a clock pendulum, a country's electrical grid, or even the ecology of plants and animals in a lake.

The second tool, called "cognitive-affective maps" or CAMs, was invented by Paul Thagard, a philosopher and cognitive scientist; it's a way of diagramming people's mental networks of emotionally charged concepts. In contrast to the state-space method, which helps us understand how distinct worldviews are situated relative to each other, CAMs give us a detailed internal view of a single worldview as it looks and feels to the individual holding it. Paul helped us understand the method, and together we were soon drawing cognitive-affective maps of the worldviews of people involved in bitter disagreements—like those over the status of the Western Wall in Jerusalem or over the oil-sands industry in Canada—while one of my doctoral students, Manjana Milkoreit, graphed the arguments over climate change. All these maps revealed hidden features of the disputes and suggested ways people might be brought together.

Later, I'll describe both tools and how to use them. Together, the state-space and CAM methods can, I believe, help us bend humanity's curve upwards again. For one thing, they can improve our understanding of our own worldviews and those of people around us, so that we can better identify potential allies and judge whether and how our differences with others can be bridged. They can also ease our fear and anger when we're dealing with people who see the world very differently from us. Finally, they can help us navigate more strategically through tomorrow's complex—and conflict-ridden—political terrains. All these benefits will, in turn, make us more effective as we engage with—and try to mobilize—other people to address our common problems. In short, they'll make solving these problems more feasible.

From my point of view, though, the two tools offer something even more exciting: we can use them to supercharge our imaginations to explore and map what I call the "mindscape." This is the geography of all possible worldviews available to human beings. In my mind's eye, the mindscape looks like a landscape on Earth—a place of lakes, mountains, oceans, valleys, rivers, plateaus, canyons, and cliffs. I imagine people using the state-space and CAM tools—as if they were compasses and sextants—to guide their journeys across this landscape as their worldviews evolve. With these devices, they're able to investigate previously hidden domains and features, and follow steep and winding paths, to discover and experiment with perspectives that they've never experienced before—some, perhaps, that no one has experienced before.

Using these tools, we can, in other words, begin to explore a vast unknown world of unseen and unrecognized worldview possibilities beyond the simulacrum of our conventional social reality. In the process, we might discover ways to catalyze the major worldview shift that our species needs to undergo if we're to have a future worth having—a shift away from resentment, anger, division, and political extremism, and towards a rough-and-ready but shared global worldview that's generous, inclusive, trusting, and compassionate—and that empowers us to live together more wisely on Earth.

# 3

# Fighting a Scarcity of Hope

*It is then in making hope practical, rather than despair convincing, that we must . . . extend our campaigns.* Raymond Williams

MY CONCERN ABOUT POLITICAL POLARIZATION and radicalization stems from my research on the sources and nature of human innovation and prosperity—research that goes back to one of my earliest university projects.

I first learned to "code" in early 1978, when I took a course in basic computer science as an undergraduate at the University of Victoria in British Columbia. I enjoyed the course immensely, particularly the way it married programming's precise logic with the latest whizbang technologies. I still remember visiting the young professor's office and marveling at the new personal computer he displayed proudly on his desk—a clunky Commodore PET, as I recall, but a true prize, because no one else at the university had one at the time.

He encouraged me to expand my programming beyond the course material, so I was soon spending hours in a windowless basement room hunched over a punch-card machine. In those days, a programmer had to convert each line of code, written in a language like FORTRAN, into a set of holes in a special piece of card stock. Once I'd made the cards, I arranged them in exactly the right sequence and slid the bundle on a tray into a glass-enclosed room that housed the university's mainframe computer—with banks of flashing lights and spinning reels of magnetic tape. A white-clad operator inserted the cards into a mechanical reader, thereby loading the program into the computer. Sometime later, I retrieved the results as a dot-matrix printout.

That 1970s mainframe was enormous, but it had a tiny fraction of the computing power of a standard smartphone today. Still, it could perform amazing mathematical gymnastics, and I was eager to put it through its paces. So I wrote away for the World3 program, and in a couple of weeks a box of nearly a thousand pre-punched cards arrived in the mail.

World3 was the computer program used in the *Limits to Growth* study, commissioned by the Club of Rome and published in 1972 by a team of MIT researchers under the direction of the famed Jay Forrester, who'd pioneered the field of system dynamics, and led by his young researcher Donella Meadows.[1] The team sought to understand how interactions between a rapidly growing human population and its industries, agriculture, and natural environment would affect human well-being through the twenty-first century. The book resulting from the study sold over thirty million copies in thirty languages, becoming one of the most famous environmental books of all time and fundamentally shaping the debate about the relationship between humanity and its material circumstances. Its message was grim: exponential growth in humanity's global economy would overshoot Earth's carrying capacity, causing economic collapse, and with it the collapse of the human population, sometime towards the middle of the twenty-first century.

I was deeply interested in the questions the Forrester team was trying to address, so with the program in hand, for long days in that basement room I traced out the causal pathways through the model's equations and learned how to tweak its variables and parameters. Then I punched out new cards, passed them to the operator, and paced the halls waiting for the printed results. Each "batch" of cards represented a different run of World3—an alternative scenario of humanity's future. In the process I developed an intimate understanding of the model, warts and all.

I learned that as a mathematical representation of the entire global human-economic-ecological system, World3 was astonishingly simple, perhaps unwisely so—with only thirty or so core equations linking variables like humanity's population size, capital investment, resource consumption, pollution, and agricultural output. Most of these equations did a good job of capturing basic physical processes—change in population size as a result of births and deaths, for instance. But I also discovered that the MIT team had made a huge number of secondary assumptions about social factors—about, for example, the links between economic growth, investment in health services, and changes in life expectancy—for which I could see no obvious empirical justification.

Still, sometimes simplicity reveals potent truths. World3 was the first serious effort to use computers to model humanity and its natural environment as a single, integrated, global system. Arriving shortly after the Apollo space program's famous photographs of Earth from the Moon—a blue-white marble sitting alone in an endless black void—it strengthened the emerging feeling that Earth was a small, tightly bounded, and fragile home that the human species was, bluntly, ruining.

In the near forty years since my liaison with that punch-card machine, I've closely followed the sometimes-bitter debate over the

*Limits to Growth* study. Critics have assailed many of World3's assumptions and simplifications, most cogently its lack of a price mechanism that could stimulate conservation and invention as resources get scarcer.[2] But others have noted that by many measures, the changes observed so far in human and environmental systems are closely tracking the predictions of the model's "standard run."[3]

And when the critics declare, as they invariably do, that the original *Limits to Growth* study has been falsified, because global economic growth has continued for nearly five decades and wholesale collapse hasn't occurred, they're flatly wrong. The standard run predicted that most key variables—food output, industrial output, and services per capita, for example—would turn sharply downwards only around now (between 2020 and 2030). It showed the human population starting to collapse only around the middle of the century. The simple fact is that we don't know if the report's predictions are fully correct, because humanity hasn't run the full planetary experiment yet.

In any case, the debate about exactly when collapse might happen is beside the point. Given the complexities of the global human-ecological system—especially of its social and political components—no computer model, regardless of how detailed, sophisticated, and well-grounded in evidence, could predict such outcomes with precision, as the study's authors noted at the time. What's not beside the point is whether we're now confronting firm limits to the growth of our global economy. And in the last fifty years we've learned something important about that topic.

Yes, limits do exist, and humanity is now starting to hit them hard and fast. But those limits are different in a critical way from those foreseen by the World3 team.

The MIT researchers distinguished between nonrenewable resources—especially metal ores and fossil-fuels reserves, including petroleum and coal—and renewables that can replenish themselves over time, like forests, fish stocks, and fertile soil. Worsening scarcity of nonrenewable resources was the main driver of crisis in the World3 model. But it turns out that the distinction between nonrenewable and renewable resources is misleading, in part because the very label "renewable" suggests that such resources are more abundant and less vulnerable to degradation than nonrenewables. Instead, the critical distinction is between what we can call "complex" and "non-complex" resources.

Nonrenewable resources like copper ore and oil aren't complex: they don't interact continually with other parts of a surrounding system; they just sit in the ground until they're extracted. And while the quality of these stocks may fall over time—as we extract the best, most accessible ores first, for instance—we can compensate for that declining quality more or less indefinitely, by investing steadily more ingenuity and energy to dig up and process progressively lower-quality ores, drill deeper for oil reserves, convert vast quantities of bitumen into diesel fuel, and the like.[4] If we're prepared to ravage every corner of the planet, scarcities of these non-complex resources probably won't limit economic growth for a long time to come. In the last fifty years, we've learned that the MIT team was simply wrong on that count.

Yet the situation is very different when it comes to complex resources—those that are dynamic, always changing systems, like Earth's climate and the planet's cycles of fresh water, nitrogen, and carbon, as well as its living systems, like fisheries, forests, and pollinators. These resources are renewable in some sense, yes, in that they can regenerate themselves over time under the right circumstances. However, the World3 researchers didn't seem to recognize that they have several other features that make them particularly vulnerable to overexploitation and degradation.

Our planet's complex resources are intricately connected by countless flows of matter and energy. It can be extremely hard to draw a line around such a resource, such as the cycle of water between land and atmosphere, to determine precisely what's part of it and what isn't. Also, some complex resources like ocean fisheries move around a lot and cross international political boundaries willy-nilly. So establishing property rights for complex resources is often difficult or even impossible, which generally isn't true for static ore deposits or oil fields. Because property rights aren't clear, no one individually has a strong monetary incentive to protect or conserve the resource. Managing it becomes a "collective action" problem: everyone has an incentive to grab what's available before others do, and no one is willing to bear the costs of stopping first.

In addition, because complex resources are connected, badly damaging one, like climate, can produce severe downstream damage to others, like the cropland and forests dependent on climate's rainfall. And finally, because the internal workings of complex resources are "nonlinear"—which, as I'll explain, means that the magnitude of a cause isn't proportional to the magnitude of its effect—these systems can "flip" from one state to a very different state in unpredictable and irreversible ways, as we've witnessed around the world, for instance, when ocean fisheries collapse as a result of pollution or overfishing.

All these factors combine to exacerbate scarcities of our planet's complex resources. And it turns out that it's *these* scarcities—especially of a stable and benign climate and its capacity to absorb our greenhouse gas emissions—that are the real limits to further economic growth. Climate change alone is already costing both rich and poor countries trillions of dollars in lost food output, infrastructure damage, lower worker productivity, more disease and the like, and these costs will escalate dramatically in coming decades.[5]

In *The Upside of Down*, I explored the social implications of this kind of change. I showed how "tectonic" stresses operating deep in human societies—demographic, environmental, energy, and economic stresses, in particular—have at pivotal moments through human history joined to bring civilizations to collapse. For instance, the decisive cause of the Roman Empire's fall was arguably a critical energy shortage due to inadequate cropland to grow the food (calories) needed for its workers, soldiers, and draft animals, which then weakened the empire's ability to maintain the complexity of its institutions and thus the well-being of its citizens.

Today, as I'll show later (in chapter 9), a somewhat different set of tectonic stresses is operating worldwide, most importantly worsening economic insecurity, mass migration, climate change, and what specialists call "normative threat," which arises when people feel their society's essential fabric of cultures, moral values, and shared beliefs, myths, and practices is being shredded. This combination of stresses is what has already led many people to feel that their situations are increasingly unsafe and unfair, helping to cause a profound shift in humanity's social mood—from a general expectation that the future will be more abundant and prosperous than the present to a feeling, still somewhat vague and unarticulated, that the future will be marked by acute scarcity.

When limits to growth really start to bite in coming years, this shift could accelerate quickly in a truly horrible direction, and as a social scientist I use the word *horrible* without exaggeration. The changes in our everyday material circumstances in the future will be harmful enough—the heat waves, droughts, wildfires, and food shortages arising from climate change, to mention just a few examples—but their impact could be multiplied a thousand-fold by flips in our emotional attitudes and worldviews.

The greatest danger is that we'll lose faith in our ability to create a positive future, and with it much of our hope. Scarcity of hope could turn out to be the most crippling scarcity of all.

How do we fight that scarcity?

### AN ANTITHESIS TO MAD MAX

Around 2015, people who follow mainstream climate science that's published in the world's leading journals like *Nature*, *Science*, and the *Proceedings of the National Academy of Sciences* noticed a shift in the tone of the reports and analyses. The overall message seemed to become much more urgent. Multiple indicators of warming's rate—the figures, for instance, on the annual increment of extra $CO_2$ in the atmosphere, the annual increase in average tropospheric temperature, the amount of heat absorbed by oceans, and the rate of meltwater discharge from Greenland—were coming in towards the more dangerous end of their previously predicted range of values. Also, scientists began to express much greater concern about run away warming kicking in at lower temperatures than previously thought possible, even at an increase of less than 2 degrees—propelled by feedbacks, as when increasingly massive forest fires release huge amounts of $CO_2$, causing more warming and fires.

In late 2019, US scientists announced that the Arctic had become a net source, rather than a sink, of carbon emissions to the atmosphere, indicating that such a vicious circle had already begun in the planet's north.[6] Meanwhile, other scientists suggested that several of the world's breadbaskets could be hit by climate shocks simultaneously, generating a crisis in food supply for the entire planet.[7]

A thoughtful colleague recently told me why social scientists who understand this latest science are usually much more worried about our collective future than even the climate scientists themselves.

"The climate scientists know the science is now really bad," he said, "but the social scientists understand the implications of that climate science for social order. They know how fragile that order is, and how quickly it can fall apart."

So, here's the sum of my fears, as a social scientist who has studied how major stresses in social systems can escalate, collide, and then swiftly multiply to produce social collapse. Sometime in the not-so-distant future, about 2030 or 2040—well after humanity has largely recovered from the economic and social damage wrought by the COVID-19 pandemic, but right about the time the World3 model indicates, perhaps coincidentally, that humanity will hit a much more unforgiving wall—a series of undeniably disastrous climate events occurs.[8] Maybe the world's boreal forests burst into flames over immense areas of Alaska, Canada, Scandinavia, and Siberia—areas orders of magnitude larger than those burning each summer now—shrouding much of the Northern Hemisphere in smoke; maybe large portions of the Antarctic ice sheet start to slide into the ocean and sea levels start rising far faster than predicted—say, a meter every ten years—causing people around the world to begin abandoning coastal zones, and property values in those zones to plummet; or maybe the South Asian monsoon fails for several consecutive years, prompting hundreds of millions of people to try to leave the desiccated subcontinent, and triggering a war, perhaps even a nuclear war, between India and Pakistan.

Whatever the events, a switch then flips in the psyches of great numbers of people everywhere, and many of us jump en masse from one worldview in the mindscape to another. The majority of us who've been trying to pretend that climate change is "fake news" or isn't too serious, or is someone else's problem, or will have real consequences only far off in the future, or can be managed when it gets serious, suddenly realize three things at once: the situation is already becoming catastrophic; everything we care about—our jobs, the well-being of our family members and communities, our homes, the very landscapes around us—is at grave risk; and, most importantly, it's *too late* to do anything significant to forestall catastrophe on a world scale. In that instant, as awareness of these three facts flashes as a meme through humanity's hyper-connected global networks, infecting our collective zeitgeist, social and institutional order starts to disintegrate as people everywhere come to one inescapable conclusion: only those ready to fight will survive.

This vision of the future as a descent into a kind of Hobbesian state of nature—a vicious zero-sum struggle of all against all as scarcity worsens; institutions, trust, and social order break down; and hope for a positive future evaporates—is already a staple of popular culture. Its visceral emotional impact is vividly conveyed by the *Mad Max* series of films, in which ex-policeman Max Rockatansky roams across a post-apocalyptic Australia encountering brutalized remnants of civilization constantly embroiled in barbaric violence.

But just because the vision is commonplace, even hackneyed, doesn't mean it's inaccurate. One way or the other, this century's volatile material, social, and psychological circumstances are going to create conditions ripe for jumps to new worldviews, as we struggle to make sense of these circumstances. The tectonic stresses I've just identified will produce pulses of social earthquakes (detailed later) that will then progressively discredit today's dominant worldviews and the institutions and technologies associated with them, including the ideology of capitalist economic growth. This process will create mental and social space for new ideas to develop, both malign and benign. Almost certainly, social turmoil, fear, and anger will drive many people to worldviews—and to "hero stories" drawn from these worldviews; that is, to stories that people use to buffer themselves against their rising death anxiety—that eschew reason and promote intolerance and violence. So a cluster of worldviews of the Hobbesian, Mad Max kind will coalesce in the mindscape.

Around the world, today's right-wing populist leaders are of course already exploiting many of these Mad Max ideas and emotions—ideas and emotions aligned with white-supremacy ideology, climate-change denial, and authoritarian nationalism—to gain huge support from segments of the general public. Yet while this specific cluster of worldviews in the mindscape seems to be coalescing fast, history, philosophy, and complexity science all tell us that something much different, and perhaps even much better, should emerge too.

Powerful worldviews tend to create their own antitheses, just like today's dominant capitalist worldview has spawned alternatives that define themselves in opposition to it, such as the Occupy movement—or like the rise of present-day right-wing populism has triggered, in response, a mobilization of support in many Western societies for liberal democratic values and institutions. From the viewpoint of complexity science, such antithetical worldviews "co-evolve," which means they reciprocally affect each other, partly by being useful vivid counterpoints to each other, but also by altering the evolutionary environment in which each survives or fails.

But we shouldn't expect this evolution to happen in coming decades as it has in the past, because humanity's current situation is simply unprecedented. Our species is confronting an interconnected set of urgent, self-generated crises that will ultimately affect *everyone* on the planet, and we're simultaneously becoming widely aware (despite obstinate and self-serving denial in some powerful quarters) of the reality of these crises and their underlying causes. So the process of worldview evolution is likely to be—it needs to be—much more conscious and intentional than before. The new combination of urgency, common fate, and emerging awareness will catalyze at least some people to actively invent alternative worldviews.[9]

As the century unfolds, fear is likely to become humanity's overriding emotion. Successful worldviews—those that survive and spread through large populations—will exploit this fear to motivate people's hero stories, for bad. . . . or just maybe for good. People's heightened anxiety about their own mortality and that of their children *could* induce broad cooperation instead of conflict, because some people will recognize that humanity has a stark choice between denial, disengagement, and division on one hand, and honesty, engagement, and unity on the other—that, put simply, it's not the case that some of us will live, while others of us will die; instead, we're either going to survive together or die together. This century's unfolding trauma could be exactly what we need to ensure we flourish together in the future, by obliging us to learn some collective wisdom and develop institutions, ways of governance, and technologies that promote opportunity, ensure fairness, and guarantee our security within the remaining fabric of life on Earth.

That's all well and good, one might protest, but if the world is burning and drowning in equal measure, how could any alternative, *humane* worldview be powerful enough to dominate in the co-evolutionary struggle with its Hobbesian, anti-democratic, Mad Max counterparts, especially in a world saturated with fear? Can we really imagine what this kind of humane worldview might be like? And even if we can, how could enough of us migrate to that worldview fast enough—how could enough of us adopt such a worldview in time—to put humanity on a genuinely better path?

To imagine and adopt alternative worldviews that are both humane and powerful, I believe we need a new attitude towards and understanding of hope. The right kind of hope and clarity about what hope should look like will motivate our agency, and that sense of agency will then make it possible for us to come to worldviews that are more humane and fruitful than the dominant ones today.

But for hope to have this motivating force, it needs an object, one that allows people to weave together, into a compelling narrative, their complementary hero stories. That object should be a positive vision of the future. And to craft a transformative vision, we need to imagine together—even to "design" together using the kind of tools I'm introducing here—alternative worldviews that provide the necessary scaffolding for that vision. We need, in other words, to create a virtuous circle—what complexity scientists call a feedback loop—linking our worldviews, our vision of the future, and our hope in both our minds and the world we're trying to change.

In the chapters ahead, I'll lay out my analysis for each of these three elements and show how they can operate together. I'll argue that we should anchor our *worldviews* in key moral truths and a strong belief in our agency, orient them towards the future, and ensure they're both receptive to change and generous to others and nature; that we should build our shared *vision of the future* around principles of opportunity, safety, justice, and a species-wide feeling of "we-ness"; and that, within the context of these worldviews and this vision, we should reinvent our *hope* so that it's honest, astute, and powerful.

## 4

# So Our Souls Can Breathe

*No man lives, can live, without having some object in view, and making efforts to attain that object. But when object there is none, and hope is entirely fled, anguish often turns a man into a monster.* Fyodor Dostoevsky

ON A SUMMER EVENING IN 2017, around the second anniversary of my father's death, I again found myself looking across Juan de Fuca Strait. Earlier, brisk westerlies had scoured the sky of the last wisps of forest-fire smoke. Now in the hour before sunset, the winds had died away, and calm had settled on the sea. The air was blessedly sweet again, and the mountains were visible at last—in fact, they were now so clearly etched against the sky that I felt I could touch them. Also gone was the dead weight of sour emotion that had accompanied the smoke.

But while the smoke had cleared, my head hadn't. Memories of summer smoke and winter rains tangled in my mind. I imagined Ben, Kate, and other children struggling as aging adults through their last years on an unrecognizable Earth, its glorious natural abundance diminished, its societies rife with conflict. That mental picture contrasted so jarringly with the vista of seemingly unharmed nature in front of me that I wanted to dismiss it as a nasty hallucination.

But I knew, again, that the best scientific evidence says such a future will come to pass if humanity continues in the direction it's going. So my mind kept returning to the basic question: What can we do to change our course for the better? I felt like I was stumbling around a room whose smooth, seamless walls—representing the various crises humanity is facing—were moving relentlessly inwards. There's very little time. Does the room have a door that will let us escape from catastrophe, and if so, how can I find it?

Hannah Arendt once wrote about "the calm that settles after all hopes have died."[1] Something about the calm sea allowed me, at that moment, to face the possibility that our hope for a recognizable future will die. The idea rested in my mind for a while. And then, deliberately and decisively, I rebelled against it. Particularly as a parent of young children, accepting hope's death is, very simply, unacceptable—far more so, even, than accepting the prospect of an unrecognizable Earth. This is a bedrock truth. And a starting point.

Bluntly, without hope, humanity's situation is unsalvageable; with it we might have a chance, even a good chance. Psychologists usually define hope as a state of mind—a person's desire or longing for an imagined set of circumstances that might occur in the future; and they've shown that loss of hope can become self-fulfilling by fatally weakening our sense of agency. When we convince ourselves there's no hope—that there's nothing we can do about a critical problem we face—we won't try to avert the problem; and, of course, the then-worsening problem gives us even better grounds to forsake hope—a damning cycle. People in some societies are now evidently in danger of this kind of downward spiral: in early 2019, for instance,

Yale University researchers reported that of Americans who think global warming is happening, only 52 percent feel at all hopeful about the problem, and a stunning 62 percent feel "very" or "moderately" helpless in response to it.[2]

So as I stood on the cliff overlooking the sea, all threads in my tangled thoughts led to one conclusion: the door out of catastrophe must be in our minds, and the key to that door must be hope. But we need more than just any old kind of hope. We need something like the kind of hope Stephanie May drew upon when, against all odds, she helped stop nuclear testing and helped keep the world from poisoning itself with radioactivity—one that's honest about the critical problems around us, astute in its understanding of ours and others' perspectives on these problems, and powerful enough to galvanize all of us with the sense of agency we need to address them effectively.

Unfortunately, though, the hopes we entertain in our minds these days are usually none of these things. Mostly, they're either false, naive, or passive.

That's largely because honesty and hope are often in tension with each other. Indeed, if we're not careful, dwelling on how terrible things may become if current trends continue can simply produce hopelessness. The more we honestly accept scientific evidence about problems like climate change, for example, the more dangerous, immense, and intractable such problems may appear, and the more overwhelmed and scared we may feel—the way I felt during those January nights listening to the rain. The problems can seem so big, while each of us taken individually seems so small, that our first reaction might be to throw up our hands and walk away in despair. In these situations, honesty about what may happen only destroys our hope and enervates us.

Because we don't like being afraid, and because we don't like losing hope and agency, just about all of us choose to sacrifice at least a little bit of honesty, at least occasionally. We lie to ourselves in many ways, some obvious, others subtle and hidden, like when we tell ourselves that our dubious dietary choices won't affect our health; or that our extra speed on the road isn't dangerous, because we're better-than-average drivers; or that we don't need to set aside funds to cover a sudden loss of income. In the same way, we might tell ourselves that the risk of devastating global warming isn't so great, perhaps by ignoring or misinterpreting key evidence. We commonly call such self-deceptions wishful thinking; they can give us false—or dishonest—hope, and so help us put the problems temporarily out of our minds.

These little lies are perfectly understandable, commonplace reactions. Researchers have found that people, groups, organizations, and societies find it enormously hard to imagine and prepare for bad outcomes. As T.S. Eliot famously wrote, "Human kind cannot bear very much reality."[3] Yet today, in the context of threats of nuclear war, an escalating climate emergency, rising authoritarianism and the like, the immediate comfort such dishonesty gives us comes at an astounding cost. Our little lies blind us to what's really happening. And just as surely as hopelessness itself, our false hope weakens our motivation to aggressively tackle our problems, and thereby weakens our agency. In today's context, lying to ourselves

about the seriousness of our situation simply makes that situation more serious.

Most of the time, though, most of us fall somewhere between the opposite poles of despair at the threats we face and hopelessness, at one extreme, and outright self-delusion and false hope, at the other. Nervously, we inhabit the middle ground, and in this middle we're deeply ambivalent about the power and usefulness of hope. We don't want to give up on it, but our honest intuition that lots of things are going wrong—our legitimate worry that the possibility of a good future for ourselves and our children is slipping away—makes the idea of hope, like our faith in human progress, seem a bit silly, even pathetic. Any hope we have for the future feels naive and passive—an expression of our deep-seated anxiety rather than an enthusiastic anticipation of the future's possibility and promise.

Some hard-nosed realists might respond, "Well, so much the worse for hope! Hope today is a meek answer to a bewildering world that's becoming increasingly unhinged and treacherous. The sensible attitude towards our present reality—and therefore towards a potentially harrowing future—should be cynical self-interest, ironic detachment, or perhaps even dissociative fantasy. It's not hope."

But based on research in psychology, social psychology, and philosophy—as well as long and deep reflection in the humanities and arts—I know this bleak conclusion is wrong. As the late, renowned American psychologist Charles R. Snyder summarized, even on a personal basis, "high-hope persons consistently fare better than their low-hope counterparts in the arenas of academics, athletics, physical health, psychological adjustment, and psychotherapy."[4] And most of us know intuitively that hope of some kind is vital to our personal well-being and to that of our communities too. We know that if we try to live without hope, we won't remotely flourish; we'll survive only in the most basic physical and psychological sense. The French philosopher Gabriel Marcel put it magnificently: "Hope is for the soul what breathing is for the living organism."[5]

Thus, we face a profound quandary: hope is vital to our well-being, but today hope is often under siege, either corrupted by our falsehoods or enfeebled by our worries. Perhaps, I thought as I looked across the sea, instead of giving up on hope, or losing it, we need to find it again, reimagine it, and reinvigorate it as a potent source of strength and guidance in a baffling and increasingly dangerous world. It should be honest, not delusional; passionate, not weak; astute, not naive; and brave, not timid. Most importantly, if we're going to avoid that downward spiral of resignation and loss of agency, it must be powerful, not passive. It must give us a real sense of purpose for positive action.

The sun had set, and the air was turning chilly. Dusk was making everything around me slightly indistinct. But as I turned away from the mountains and sea, my mind was clearer. The evening's calm had taken me beyond the despair of hope's death to something potentially even thrilling. I could glimpse the possibility of a new and better path forward for our children—and for the rest of us too.

My journey to realizing that hope was key had taken me through three successive questions. I'd begun with: What story should we tell our kids and ourselves about the future? This question highlights our need for a shared vision of a positive future. It also pointed me to a second question: What can we do to change humanity's present course to realize that vision? This question, as we face surging injustice, political extremism, and social unrest, underscores our need for radical advances in understanding each other's worldviews. Then, the third and equally important question: How can we be honest about the extent of our problems without destroying the hope we need to sustain both our psychological well-being and our sense of agency? How, in short, do we reconcile honesty and hope?

The last question underlines a profound moral and psychological challenge we must all face if we're to avoid the trap of false hope: doing the moral thing—being truly honest with ourselves about the world's problems—can sap our agency; yet without agency, even if we can tell a positive story about the future, we won't be able to find our way to that future. Our vision and understanding must be coupled with our motivation and capacity to act.

In the months after I found myself looking across Juan de Fuca Strait, I explored diverse fields to answer these three questions and to learn about every aspect of hope. I learned how the meaning of hope has changed over time and how it differs subtly across cultures. I found that people won't hope for something they believe is guaranteed to happen and yet can't hope for something they think is impossible. And I discovered the key difference between "hope that," which is a passive and timid locution, and "hope to," which is active and bold—a difference that bears crucially on the issue of our agency as we try to deal with humanity's problems.

# 5

# The Ways Hope Works

*Despair is only for those who see*
*the end beyond all doubt.* Gandalf

SIX DECADES OF COMPRESSION and intermittent dampness had resealed yet another envelope, I saw. It was attached facedown in a weathered scrapbook open on the coffee table in front of me, its light-blue back showing no return address or other identifying information. I took a slim paring knife and separated the tongue from the envelope's main body, trying not to tear the fragile airmail paper. Lifting the tongue gently out of the way, I slid out the letter inside—a single sheet, neatly folded in four—and opened it.

The sender's address was centered at the top and printed in scarlet. The date, typed manually, appeared immediately underneath: 29th December, 1961.

From: The Earl Russell, O.M., F.R.S.
Plas Penrhyn, Penrhyndeudraeth, Merioneth,
Tel. Penrhyndeudraeth 242.

Mrs. John May,
Duncaster Road,
Bloomfield, Conn.,
U.S.A.

Dear Mrs. May,

Thank you for your letter, which I found most stimulating. I agree with you that the last year has been a dangerous and disillusioning time for us all. In a period such as our own we cannot ask for certainties: there are none. We are faced with the alternative of acquiescing in the drift to suicide, or of trying to prevent it. We betray our own humanity if we do not make the attempt. I am sure that we are mistaken if we think that the great mass of people are not interested in the problems that concern us so much. People are concerned, but they feel in a state of such helplessness that their only answer is to refuse to think about the terrors of the arms race. I am convinced that the prevalent apathy which is the only condition under which Kruschev and Kennedy, Macmillan and Adenauer, de Gaulle and Mao Tse Tung, are able to carry out their lunatic policies, is based not on a lack of concern but on a sense of impotence. If we can show people a way in which they can genuinely obstruct, and finally prevent, the whole nuclear policy, this sense of powerlessness will go, and with it the apathy. We are seeing the truth of this in the astonishing growth of the Committee of 100 in Britain.

Yours sincerely,
Bertrand Russell

I was flabbergasted. In my hands was a personal note from one of my heroes—Bertrand Russell, a philosopher and polymath of prodigious intellect and one of history's great activists for peace. But what struck me more was the ring of Russell's words: although decades old, they sounded like they'd been written yesterday. They spoke of "the drift to suicide," of the betrayal of "our own humanity" if we don't try to stop that drift, and of the apathy and inaction that arise from a "sense of impotence." These phrases—typed in the dark months after the collapse of the moratorium on nuclear testing—seemed even more applicable in the spring of 2018 than in the winter of 1961.

I'd found the letter in one of two huge scrapbooks. Stephanie May had put them together evidently as a meticulous record of her project to stop nuclear weapons testing in the fifties and of her political activism in the sixties. My exploration of hope had led me to Stephanie, and tracing her story had taken me to a log cabin in the remote and rugged Margaree Valley of Cape Breton, on the exact opposite edge of North America from Vancouver Island. In the early 1970s, Stephanie and her husband, John, had moved their young family to this distant region, completely disengaging from their previous lives in the United States. (They continued to do battle as environmental activists, though, and were later key in stopping the aerial spraying of Cape Breton's forests with Agent Orange.)

Serendipity had also played a key role in my journey to Cape Breton. When I'd started at the University of Victoria in the mid-seventies, I'd been fortunate to have as a mentor Bill Epstein, the senior Canadian in the United Nations Secretariat and a leading expert on preventing the spread of nuclear weapons to non-nuclear states. Bill was on leave from the UN to write a book, and one day he'd remarked to me, "You know, it was mothers who stopped the testing; they organized huge protests around the world, and in the end the leaders of the nuclear powers couldn't ignore them."

I'd filed that information in the back of my mind. Forty years later, thinking about how everyday people might use hope to change the course of history, I recalled Bill's words and decided to investigate further. I found online a short, unattributed passage with the intriguing subtitle "My mother stopped global nuclear weapons testing." After digging, I discovered the author was Elizabeth May—Stephanie May's daughter, the leader of Canada's Green Party, and, remarkably, someone I already knew. I wrote to Elizabeth to learn more, and she generously introduced me to her brother, Geoffrey, now in his sixties and still living in Margaree Valley.

When I arrived in the valley a few weeks later, Geoffrey soon dropped by the cabin I'd rented. Warm, funny, and wonderfully loquacious, he brought with him five hundred typed pages of the only copy of his mother's unpublished memoirs. This document was a main reason I'd come, so I was eager to start reading. But then he laid on the table two large packages wrapped in plastic. "I think you'll be interested in these scrapbooks too."

The instant I opened the first, I understood his remark. History burst to life from its pages, via hundreds of photographs, telegrams, newspaper clippings, and technical articles about some of the most noteworthy events of the mid-twentieth century. The scrapbook also contained original correspondence, much deeply personal and revealing, with people from all walks of life from around the world, including some of the most influential people on Earth at the time.

Other than Elizabeth's two brief accounts of her mother's efforts, one of which I'd found online, very little information about Stephanie's story was publicly available. Only as I got deeper into the materials Geoffrey had brought me did I realize how extraordinary the story was. Her project to stop nuclear weapons testing—in its creativity, strategy, scope, and effectiveness—was a case study in brilliant political activism. And the kind of hope that strengthened

and motivated her activism displayed exactly the features that we all need in our hope today—features that are, in fact, common to the hope of history's most effective political activists, from Mahatma Gandhi and Nelson Mandela to Greta Thunberg, the young Swede who in 2018 triggered the worldwide School Strike movement to protest the climate crisis.

For instance, Stephanie's hope was honest—in three vitally important ways:

1. It didn't hide from the danger of nuclear testing, but instead directly studied, faced, and acknowledged that danger.
2. It never rested on exaggeration or selective use of evidence. She knew the technical aspects of nuclear testing inside and out. "I read the newspapers with a pair of scissors in my hand," she wrote in her memoirs. "When I found an item about the latest scientific evidence on fallout, I clipped it and pasted it in a nuclear information scrapbook, which I could refer to when I wrote letters."[1] She also read voluminous reports on the topic of testing from Congress and government agencies. So she could rebut in detail the arguments advanced by the head of the Atomic Energy Commission, who had written to her at President Eisenhower's request. But when experts differed on testing's danger, as they did on the number of people already killed by testing's fallout, she

referred in her campaigns to the expert view that downplayed the danger the most—saying, in effect, that even the least-worst outcome was bad enough.

3. And her hope was grounded in an acute moral clarity, which translated into an unerring ability to discern the moral idiocy of the positions taken by many testing advocates, people often deeply compromised by their bureaucratic, political, or business interests in testing. When aggressively challenged, as she often was, she could return to that firm, honest footing.

Her hope was astute, too, leveraging her savvy understanding of her allies' and opponents' motivations and worldviews. Her scrapbooks and memoirs reveal how she deliberately worked with the grain of the time's popular mood. She also appealed to people's better nature by starting from the assumption that her opponents were good people who wanted to do what was right; she used, to great effect, her status as a housewife and others' ready assumption that she was naive; and she had a keen sense for the force of argument, searching out and exploiting the critical assumption, fact, or contradiction on which everything at that moment hinged.

And finally, her hope was powerful, backed by a combination of gumption, dogged perseverance, and exacting attention to political strategy, all in service of a clear vision of a better world in which all peoples were united around common principles. She knew precisely where she wanted to go, and she thought carefully about how to get there, given the resources available to her and her allies. Her strategic thinking and intuitions, as is clear in her memoirs and scrapbooks, were often far superior to those of the men and women with whom she collaborated, many vastly more professionally elevated and supposedly better informed about the world's ways. Coupled with her courage and determination, her strategic sense made her a terrifically powerful agent of change.

Yet I also learned from the scrapbooks that there were times when even Stephanie's formidable hope appeared to have been pushed to its very limits. December 1961 was one such time. By then, she'd corresponded with "Bertie"—now Lord Bertrand Russell—who was nearly ninety, for several years. She'd also met him in person at his home in Wales after the 1960 Aldermaston march; those were days of promise and optimism about the cause. But little more than a year later, nuclear testing had resumed. And by the time Russell wrote that letter I discovered, after Stephanie had devoted five exhausting years to stopping testing, the cause seemed virtually lost.

Still, on close reading, I thought the letter held a clue as to what kept hope alive for people like Bertrand Russell and Stephanie in those grim days. "In a period such as our own we cannot ask for certainties: there are none." At first blush, Russell seems to be saying that their cause is worth pursuing, not because of any prospect of success, but simply because it's morally right. Yet perhaps something additional was on his mind. Perhaps he was trying to tell Stephanie that the absence of certainty can, itself, be a reason for hope.

### IS IT TOO LATE?

Any time I'm tempted to give up on the future, I recall a vivid moment one evening some years ago, when I was traveling to talk about a book I'd co-edited on humanity's climate and energy problems. I was sitting on a small stage with the author of one of the book's chapters, before a hall packed with people. We'd had a lively exchange with the audience for over an hour, when a middle-aged man near the front interjected:

"Is it too late? Is it hopeless now?"

For a few moments, there wasn't a sound. The question brought to a fine point the entire evening's conversation. Everyone understood its unstated premise.

Whether a situation has moved past the point of no return and is therefore hopeless—truly never-to-be-recovered-from hopeless—partly depends on what we hope will come to pass—that is, on our hope's object. For instance, we're not past the point of no return, if we're hoping that some kind of life will continue on Earth. Even if humanity produces, as the science predicts, a sixth great extinction this century and next—one that eliminates from half to 95 percent of all species—life itself will likely survive in multiple forms (although it could take tens of millions of years for Earth's panorama of species to recover to today's levels, if it ever does). Or if we're merely hoping that our own species will persist, we can be sure that somewhere some people will survive problems like the climate crisis, given that homo sapiens are incredibly adaptable and resilient.[2]

But it's plausibly too late if we're hoping that human civilization this century and beyond will be just, peaceful, and prosperous. "Yes," my fellow author replied gently. "I think it is too late." The difficulty, he argued, of changing our energy systems, people's appetites for material things, the commitment of governments worldwide to endless economic growth that gobbles resources and spews out waste, and damage from climate change ensured that by the end of this century societies will be far poorer, more violent, and less free. In fact, he concluded, the widespread collapse of human civilization is a real possibility.

On the surface, I couldn't find a lot to dispute in his answer. After all, when it comes to humanity's prospects, I'm not generally known as a fount of feel-good optimism. But all the same, something in my colleague's fair response didn't seem quite right. In light of my two decades' study of the behavior of complex systems,

I felt his answer implied an omniscience we simply don't have. When it came time for me to answer, I hesitated for a moment and then said:

"I agree . . . mostly. But I'm not sure that we can say definitively it's too late—or hopeless—because we simply can't predict the future behavior of the systems that we're part of accurately enough to know one way or the other."

People didn't seem very reassured, so I decided I'd better elaborate with an anecdote. When I was in my mid-twenties, I recounted, I'd hitchhiked all over South Africa with a friend. It was the early 1980s, and apartheid was still in full force—communities, trains and buses, beaches and washrooms were segregated by race; racial intermarriage was illegal; and Blacks by the tens of thousands were still being herded, often violently, into impoverished rural "homelands." During our travels, we spoke at length to people from every racial group, and sometimes were invited to stay with them in their homes. And they all told us the same thing: apartheid is going to end, soon, but the end will bring massive bloodshed. To folks on the ground at the time, racial civil war seemed inevitable.

Apartheid did end, and while unhappily much blood was shed, the nationwide bloodbath didn't happen. Not only was a full-scale civil war averted, but a half-dozen years after our travels in the country, a white president and once-ardent supporter of apartheid, F.W. de Klerk, released Nelson Mandela from prison and then worked closely with him to negotiate and implement a peaceful transition to a multiracial democracy. In the early 1980s, nobody imagined such an outcome; and if one had described the possibility, it would have been derided as a fairytale.

By the time of our evening discussion, post-apartheid South Africa had lost much of its early luster and promise. Everyone in the room understood that the country was becoming once again a deeply troubled society and corrupted democracy—although this time, at least, nominally multiracial. But my anecdote did reach many of the people there: one rarely knows enough about the internal workings of social systems, or the external factors influencing those systems, to say that all futures worthy of hope are off limits—or beyond the boundary of reasonable possibility.

The very complexity of our social systems can preserve possibility, and our awareness of that possibility can sustain hope.

## COMPLEXITY'S VEIL

Even a relatively simple system like a wind-up clock has three basic constituents: a set of parts or components, a persistent pattern of causal links between those parts, and a flow of energy through and between those parts that sustains the pattern of links.[3] But complex systems—whether a rainforest, a human brain, Earth's climate, the Roman Empire, the global financial system, or a modern country such as South Africa—share a cluster of other features, too. They usually have a truly immense number of parts, and those parts are of many kinds—for example, all the species of plants and animals in a rainforest, from bacteria and nematodes to birds and trees. Also, the causal links among a complex system's parts typically assume many forms and can change over time. In a rainforest, relationships among species, including the flows of elements like carbon and nitrogen between organisms, are constantly reconfiguring themselves.

Together, a complex system's multitude of parts and its changing patterns of causation mean that the number of ways its parts might combine and interact together can be staggeringly large. These combinations can be a powerful source of innovation. Imagine the diverse concoctions we might create if we were to randomly mix together the components of a huge chemistry set. Usually, not much interesting would happen. But sometimes the results would be hair-raising. And occasionally, by combining the set's components in just the right way and applying a flow of energy, we might create something new and unexpected. Life itself, after all, came from such an interaction of chemicals; and it's an outcome that, as far as we know, is rare in the universe.

The interaction of a complex system's parts generates "emergent" properties and behaviors. This means we can't understand the system's properties or behaviors just by dismantling it to examine its parts separately. By this view, one can't understand life by looking only at the chemical and physical properties of its constituent molecules, like DNA; or human consciousness by looking only at properties of the brain's neurons; or South Africa's racial divisions by looking only at the attitudes of its individual citizens. It's only when the system's parts come together and interact that such features appear. The everyday adage "the whole is more than the sum of its parts" roughly captures this idea, although it's more accurate to say "the whole is *different* from the sum of its parts."[4]

Complex systems also have another key feature: feedback loops. A change in one part of the system can cause a change in another part and then another and another—in a chain that eventually either reinforces (a "positive" feedback) or dampens (a "negative" feedback) the original change. When people talk about "virtuous" or "vicious" circles, they're talking about positive feedbacks. But "positive," in this case, doesn't mean good. Think of a vicious circle where an employee is openly angry with his boss, which makes the boss treat him worse, which then makes him even angrier. This is a "positive" feedback, not because it's good—obviously in this example it isn't—but because it's self-reinforcing.

If positive feedbacks tend to reinforce change and thus destabilize systems, negative feedbacks tend to stabilize systems, by counteracting the original change or perturbation. In the above example, the initial perturbation is the employee's open anger with his boss. But if the boss is a reflective person, and the employee's anger triggers her to change her behavior towards him in a way that lessens his anger, then the causal sequence becomes a negative feedback, stabilizing the situation.

Understanding how a complex system works often involves identifying these feedbacks and determining whether positive ones dominate negative ones, or vice versa. For instance, scientists were once uncertain whether positive or negative feedbacks predominated in the climate system. But now they're confident that positive feedbacks—as when warming causes wildfires that release carbon that causes more warming—are on balance much more powerful, which is why they're now so worried about the prospect of runaway warming.

Together, lots of interactions and feedbacks among the parts of a system produce what complexity scientists call "disproportionate causation." This means there's often no clear relationship in the system between the size of a cause and the size of its effect. Sometimes, a small change in the system might produce an enormous effect—a final straw might break a camel's back or, as the Monty Python troupe joked, that last wafer-thin mint might cause a mighty gastric explosion. But other times, even a very large change in the system could produce little effect overall.

Simple machines, in contrast, don't have lots of interactions and feedbacks among their parts, so small changes usually produce small effects, and large changes usually produce large effects. If one winds a mechanical clock a little, it runs only a short period of time; if one winds it a bit more, it runs proportionately longer.

We've all heard the famous saying about Earth's weather—that the flap of a butterfly wing in one place can cause a cascade of ever-growing atmospheric changes that might eventually generate a hurricane somewhere far away. This kind of sensitivity to small

events can generate "chaos," in which the behavior of a system, like Earth's weather, looks, at least on the surface, completely random and thus unpredictable. But it's generally not appreciated that in complex systems the opposite can also be true: sometimes an event even as large, metaphorically, as a hurricane leaves the system in which it has occurred looking and behaving pretty much as it did before. And to make the challenge of prediction even harder, one often can't be sure in advance which small events will matter a lot and which big ones won't.

This disproportionate causation is what complexity scientists mean when they talk about "nonlinearity." It's one reason why complex systems like Earth's climate, forests, and fisheries, and our global financial systems sometimes shift or flip unexpectedly from one stable state or "equilibrium" to another (as highlighted in the scientific article that Kate, when she was four, found on Sarah's desk). Given the current state of complexity science, we usually can't predict the timing of these sudden shifts with any precision; we may know that the likelihood of a flip in a specific system is rising—that the probability of an earthquake, a shift in monsoon patterns, a pandemic of infectious disease, or a major financial crisis is going up, for example—but have no clear idea when the event will actually occur. Also, sudden shifts are usually extremely hard to reverse. If a fishery collapses, Earth's climate reorganizes itself, or a population's predominant worldview shifts, we almost certainly won't be able to push the system back to its old state, even if we completely turn back the original conditions that caused the flip in the first place. In such cases, complexity scientists say the system exhibits "hysteresis."

Over the years, I've found these complex-systems ideas tremendously helpful, and they were in the back of my mind when I answered the tough question that evening about hope. They suggest that uncertainty is a more important factor than is generally recognized when it comes to anticipating how social systems like economies, governments, and popular movements will behave.[5] Changes in the systems' fine-grained details can cause dramatic differences in how things turn out. And since, at any time, we can grasp only a tiny fraction of all the fine-grained details—and we know even less about how changes in those details might make a big difference—we should avoid making definitive and sweeping statements like "it's too late" or "it's hopeless."[6]

More subtly, but I think most critically, when we acknowledge the complexity of our reality, we're less likely to assume that uncertainty is always dangerous and that surprises are always bad. We're less likely, as the American writer Rebecca Solnit vividly expressed it, to "transform the future's unknowability into . . . the fulfillment of all our dread, the place beyond which there is no way forward."[7] While uncertainty can quite reasonably provoke fear—fear of the unknown—it can also give us grounds for hope, because it creates a mental space in which we can imagine positive possibilities. Sometimes the future's unknowns, when they finally reveal themselves to us, are beneficial, and sometimes surprises turn out to be good, as happened in South Africa in the late 1980s.

So, if Bertrand Russell meant to suggest to Stephanie that they could take some solace from the absence of certainties, he was on solid ground. Hope only exists when there's uncertainty. In the words of Joseph Godfrey, an American philosopher of religion, it's found "where the future is veiled."[8]

### SOLVING THE HOPE PUZZLE

Hope is a true puzzle, and this link between hope and uncertainty is one piece of that puzzle. But any one piece by itself tells us little without a context of the other pieces assembled into the full picture. Assembling that picture is remarkably difficult, partly because nearly everyone is a practical expert on the subject of hope.

We've all possessed hope, desired it, felt its loss, and seen all these things in our families, friends, and other people. These intimate experiences make hope highly contentious: some of us treasure and claim it as the bedrock of our lives; others, especially these days, are ambivalent about or even contemptuous of it. Most of us, as adults, probably hold more than one view of hope at the same time. Indeed, hope seems to be one of those vital problems "of the human heart in conflict with itself" that the great writer William Faulkner spoke of decades ago.[9]

We can find fascinating evidence of this struggle in, of all places, *Bartlett's Familiar Quotations*. My dog-eared edition of the classic reference work has hundreds of quotations incorporating the word *hope*, yet there's barely a shred of agreement among them about hope's nature or value. I extracted thirty-four particularly interesting ones and found that they split almost evenly between those expressing a negative attitude towards hope and those expressing a positive attitude, with a few "neutrals" in between. They ran the gamut from the Roman playwright Terence's neutral observation that "Where there's life, there's hope" (mistakenly attributed to Cicero in my edition of *Bartlett's*) to Benjamin Franklin's admonition that "He that lives upon hope will die fasting" and on to Emily Dickinson's effusive verse:

> "Hope" is the thing with feathers—
> That perches in the soul—

And sings the tune without the words—
And never stops—at all—

Our often-conflicted attitude towards hope isn't a modern phenomenon either: it can be traced back to ancient times—to history's most famous parable about hope, Pandora's box. As recounted by the ancient Greek didactic poet Hesiod around 700 BCE, Zeus was angry that humanity had obtained fire, so he fashioned for men the first woman, the beautiful and beguiling Pandora. She brought to humanity a jar—the Dutch Renaissance scholar Erasmus mistranslated the ancient Greek word *pithos* as "box"—that she then opened, releasing into the world a host of evils like toil, sickness, and death. But not everything escaped:

Hope only did not fly. She stayed behind
In her impregnable home beneath the lip
Of the jar; before she had a chance to slip
Out, woman closed the lid, as Zeus designed.[10]

The story is misogynistic, but what it implies about hope is intriguing. For the ancient Greeks, hope was the personified spirit, or daemon, Elpis. She carried a bundle of positive and negative connotations, some like our modern understanding of hope but others resembling today's *expectation* and *foreboding*. Classicists and other scholars have debated back and forth intensely whether the fact that Elpis stayed trapped in the jar was intended as a boon or bane for humanity, an eternal gift left behind to ease the pain of the escaped ills or, maybe, a perpetually taunting source of illusion and emotional trauma. My guess is that the parable is saying that hope is both: the ancient Greeks—or Hesiod, at least—understood that hope is ambiguous in its very essence.

Today, this ambiguity arises partly from disagreement about whether hope should have an "object"—that is, a clearly envisioned future event, situation, or state that we want to come to pass.[11] The philosophers Gabriel Marcel, Joseph Godfrey, and Jonathan Lear, for instance, all argue that hope can thrive without an object. Instead, they note, it can be an unfocused disposition—a general feeling of hopefulness or an intuition that, as Jonathan Lear puts it, "something good will emerge." He introduces what he calls "radical hope," which is "directed toward a future goodness that transcends the current ability to understand what it is." Godfrey speaks of "fundamental hope," which exhibits "an openness of spirit with respect to the future."[12]

This kind of "objectless" hope is common: we all know people who sustain a fundamentally hopeful demeanor, come what may and without a specific future state in mind. Sometimes the attitude simply rests on generic optimism, which is a general tendency to appraise future circumstances through rose-colored glasses, highlighting information that boosts the perceived likelihood of desirable outcomes and that reduces the perceived likelihood of undesirable ones. But generic optimism is often little more than wishful thinking; and it engenders false hope that can, as I suggested in the last chapter, leave us unprepared for a treacherous future.

A more interesting objectless hope rests on a firm conviction that something like a higher power or a compelling moral, normative, or spiritual framework will guide and protect us and give our lives meaning. We might tell ourselves that if we follow God's direction or live by certain ethical principles, things will ultimately work out well, whatever shape they take. Barack Obama's notion of hope seems to be of this form. In his famous 2004 keynote speech to the US Democratic National Convention, which propelled the future-president to national prominence, Obama said:

> I'm not talking about blind optimism here—the almost willful ignorance that thinks unemployment will go away if we just don't think about it, or the healthcare crisis will solve itself if we just ignore it.
>
> That's not what I'm talking about. I'm talking about something more substantial. It's the hope of slaves sitting around a fire singing freedom songs; the hope of immigrants setting out for distant shores; the hope of a young naval lieutenant bravely patrolling the Mekong Delta; the hope of a mill worker's son who dares to defy the odds; the hope of a skinny kid with a funny name who believes that America has a place for him, too.
>
> Hope . . . hope in the face of difficulty. Hope in the face of uncertainty. The audacity of hope.
>
> In the end, that is God's greatest gift to us, the bedrock of this nation; a belief in things not seen; a belief that there are better days ahead.[13]

Throughout his political life, Obama articulated a detailed policy vision of what he saw as a better future for his country, and this vision could be taken as the object of his hope. Still, the hope he actually talked about in his speeches and writings seemed to be mainly dispositional—a general psychological attitude resting on the conviction that adopting and following the liberal American creed will ensure "better days ahead."

Such convictions have great merit. They give people and groups the psychological strength to persevere through prolonged periods of staggering hardship. One thinks, for instance, of the Jewish people's perseverance through the centuries following the destruction of the Temple in Jerusalem in 70 CE. As my colleague Fred Bird writes, Jewish leaders "cultivated hopes not through expectations of a new and better future but by developing institutions, practices, literacy, and most importantly, a mindset that enabled Jews to live in the present."[14]

But, today, humanity must do far more than get through this century more or less intact. Mere perseverance isn't enough: we need to become engaged—as active agents—in addressing our monumental challenges. Objectless hope might inspire and excite us for a time—the way Obama's invocation of hope excited his audiences—but it's less likely to motivate us to lean aggressively into our problems to solve them, because by definition it doesn't offer a clearly defined alternative to those problems. It's also vulnerable to ridicule—something the former vice-presidential candidate Sarah Palin highlighted in 2010, when she asked of President Obama's program: "How's that hopey-changey stuff working out for ya?"[15] In the end, as Palin sensed, objectless hope can be caricatured as vague and naive—and as a weak reed in times of great peril.

I think the nineteenth-century English Romantic poet Samuel Taylor Coleridge had it exactly right when he said that "hope without an object cannot live." Hope gains confidence from a compelling story about what we want to come to pass. The hope we need—a commanding hope—must have a clear, motivating vision of a desirable future, if it's to be powerful and sustainable over time.

Research in analytical philosophy (including its offshoot, philosophy of mind) and positive psychology is helpful here. Both approaches usually adopt the definition of hope I gave earlier: it's a psychological state (in an individual mind) of desire or longing for an imagined set of circumstances, a possible world, that might occur in the future. Analytical philosophers study this psychological state by scrutinizing the concept of hope, which usually means identifying what beliefs and emotions a person must logically have to experience hope. In contrast, positive psychologists are interested in healthy human cognition, so they generally study real people in social and laboratory settings to understand hope's causes, forms, and effects.

Both approaches recognize that imagination is a mental process that's crucial to hope. As we'll see in the next chapter, imagination creates the visions of future possibility that, when taken as objects, act like oxygen for our hope. But given the state of the world today, imagining a motivating vision without resort to optimism's rose-colored glasses can be exceptionally hard.

Perhaps it's surprising, then, that I believe commanding hope should explicitly eschew the generic optimism I described above, including the kind of "negotiation" with reality that positive psychologists sometimes recommend, where we selectively emphasize some of its good elements and downplay others. If we're to face our future with any chance of success, our vision of the future must be anchored in clear-eyed realism. We must use the best-available knowledge to discern the true nature of the crises we're facing—including climate change, pandemics, economic distress, and political extremism—as well as the factors that will shape our collective future and limit the range of desirable worlds available to us tomorrow. While fully acknowledging uncertainty, and even seeing space for hope in that uncertainty, commanding hope must remain brutally honest about what's genuinely possible going forward.[16]

### FROM HOPE THAT TO HOPE TO

Once our hope is focused on a story about the future that's anchored in clear-eyed realism, we need to perform a bit of magic—to turn our hope "water" into hope "wine," so to speak. We need to turn our hope THAT the desired future will happen into a hope TO make it happen. Such a psychological shift makes us active agents in the story—converting us from spectators to protagonists with a role to play. It also converts the object of our hope—our vision of a desirable future—into a set of definite goals that we can then strive to achieve.[17]

We use "hope that" statements all the time in everyday conversation, as when we say: "I hope that it will be sunny tomorrow." Such statements strongly imply we're passive and have little if any control over the occurrence of the future we desire. We don't control whether it will be sunny tomorrow. The implication is much different, though, when we use "hope to" statements. We usually follow the phrase with a verb, suggesting immediately that we're active actors, that we're going to try to create the desired future, and that we think we have at least some control over that outcome. When we say, for example, "I hope to plant my vegetable garden tomorrow," we're saying we believe we have some control over whether the garden gets planted tomorrow, but we're uncertain about how much control, perhaps because the weather might be poor.[18]

Analytical philosophers generally use the "hope that" locution in their research, because they see hope as a mental attitude towards a possible state of the world (the hope's object), but that possible state doesn't have to be a goal—it can be simply an outcome that's passively desired. Positive psychologists, on the other hand, like the "hope to" locution, because they focus on how people perceive their goals, their pathways to those goals, and their capacities to follow those pathways. (For these researchers, a "pathway" is a visualized sequence of actions to reach a goal.) They focus on people's perceptions of self-efficacy—on whether people see their fate as determined by external circumstances or internal choices. They argue we can't hope TO shape our fate, or bring about our desired future, without some belief in our agency; this kind of hope needs belief in agency as a precursor.

And "hope to" can also motivate our agency. It can keep us going, by encouraging us to persist in efforts to bring some part of the world around us under our control. As Princeton philosopher Philip Pettit says: "[When] people do not despair under bad news or in evil times—when they manage to keep their hearts up and press on in positive ways—they often succeed in overcoming obstacles that might otherwise have brought them down."[19] Critically, then, when we "hope to" bring about a desired future, our agency has a dual role: it's both an input to, and an output of, our hope. In complex-systems language this dual role can create self-reinforcing, positive feedbacks—both virtuous and vicious circles—in our mental mechanisms of hope.

In the virtuous-circle version of the feedback loop, our memory of having successfully achieved something we desired in the past causes us to believe we have some agency in the present. This sense of agency then becomes an input to our present hope, which encourages us to vigorously exercise our agency to produce something we desire in the future. Experience of new success, in turn, creates another memory that's an input to our hope in the future. For instance, people who've found jobs they like in the past are generally hopeful about finding good jobs in the future; that hope contributes to their success in finding such jobs, which then further strengthens their future hope.

In the vicious circle, though, our memory of situations of sharply limited agency in the past creates weak hope in the present, which causes us to forgo trying to produce things we desire in the future—an outcome that only lays down even more memories of limited agency, and further weakens our future hope. People who have been chronically unemployed often feel hopeless about their job prospects, and that hopelessness only makes their future unemployment more likely, along with even worse hopelessness. This is the "damning cycle" of self-fulfilling loss of hope I mentioned before.

## COMMANDING HOPE

Commanding hope, as I conceive of it, is a "hope to" kind of hope: we hope TO bring about a specific vision of the future by setting out goals for achieving that vision and by making ourselves active, powerful protagonists in that vision's story. It's designed to shift us from a vicious to a virtuous circle of agency. In his letter to Stephanie May, Bertrand Russell alluded to the vital importance of such a shift when he said:

> I am convinced that the prevalent apathy . . . is based not on a lack of concern but on a sense of impotence. If we can show people a way in which they can genuinely obstruct, and finally prevent, the whole nuclear policy, this sense of powerlessness will go, and with it the apathy.

Commanding hope has three components—honest, astute, and powerful hope—that combine to help create this virtuous circle. Each has a fundamentally distinctive character and highlights a facet of the hope we need. Honest hope is a *moral* attitude, because it starts from a presumption about the moral importance of a commitment to truth. Astute hope is an *epistemological* attitude, because it's grounded in deep knowledge of people's worldviews and motivations. And powerful hope is a *psychological* attitude, because it emphasizes how a vision of a positive future and a clear roadmap of strategies to get there can motivate agency.

Over the millennia, scholars have investigated hope's moral, epistemological, and psychological features in enormous depth. But treatments of these features have often been siloed, partly, I think, because many of these thinkers haven't fully acknowledged the multifaceted nature of hope. Positive psychologists, for instance, study hope's motivational properties; but even when these psychologists are being prescriptive—when they're advising patients, for example—they tend to neglect hope's moral features. Instead, they seem inclined to strictly separate what they see as hope's scientific and descriptive "is" from its moral and prescriptive "ought."

But having reviewed what many thoughtful people have said about hope, I'm struck that moral, epistemological, and psychological features are all part of its basic nature. Any conception of hope that focuses on only one or two of the three is impoverished. So, I've explicitly integrated all three into my notion of commanding hope.

**Honest Hope**

The foundation of commanding hope is honest hope. It has the courage to fully acknowledge the dangers we face, so it's informed by a thorough scientific understanding of those dangers and the likelihood of stark constraints in our future; yet it also welcomes the possibility of genuinely positive alternatives within those constraints.

Honest hope starts by assuming that we inhabit a discernable reality; that some claims about this reality are more accurate, useful, or "true" than others; and that we can make such judgments by referring to empirical evidence and by applying the scientific method—that is, by using evidence from reality to rigorously challenge our theories and our claims about what is true. In short, honest hope regards "truth" as a scientifically meaningful and defensible concept. Truth is, of course, a ferociously contested idea these days—and one that's fundamentally tied to issues of trust. Since any one of us can know only little bits of reality directly, we must rely on other people and institutions for information about most "facts" about our world, and any commitment to truth implicitly involves trust in those people and institutions as sources of fact.

Honest hope then adopts an explicit moral stance towards truth: it holds that trying to see the truth and accepting its full implications are morally good behaviors. This stance is justified by a simple, practical argument: while sometimes shading the truth—selecting or distorting data, for instance, to make things look more positive—might help sustain our hope and agency, in the context of today's severe global problems doing these things is spectacularly counterproductive. The costs of poor preparation and engagement with these problems will far outweigh any benefits we gain from lying to ourselves about them.

That's one of the most powerful rejoinders to today's populist authoritarianism—including that of US president Donald Trump and his right-wing and corporate apologists and enablers. Deliberate, willful ignorance about the way our world works, whether it involves burying scientific data about climate change, downplaying the menace of an emerging pandemic, or removing credible economic experts from key policymaking roles, ultimately only hurts the very same discontented groups whose interests these populists say they're defending.

And the truth, even if it's harsh, doesn't have to destroy our hope, enervate us, or undermine our agency. It can invigorate our agency. Telling the truth can summon people "to a level of extraordinary greatness appropriate to an extraordinarily dangerous time," notes the American environmental thinker David Orr.[20] He cites as examples Abraham Lincoln's appeals against slavery, Winston Churchill's calls to the British people to defeat Nazism, and people everywhere facing and overcoming life-threatening illness.

**Astute Hope**

Astute hope is strategically smart. It's centered on the idea that we'll be more successful on our pathway to our desired future if we can develop an understanding of the minds of the diverse people we encounter along the way—be they friends, potential allies, or implacable opponents. Then we'll better grasp what might motivate them in response to the future's dangers, constraints, and possibilities and how our respective worldviews and motivations might mesh or clash. The tools my colleagues and I have invented to map worldviews, which I'll describe later, should make generating this kind of knowledge easier.

Our deep understanding of people's worldviews and motivations can reveal how other people see us, and how they think we see them. We can use this information to make better judgments about how our actions are likely to be perceived by others, how other people will likely respond, and how we can adjust our actions in anticipation of those likely responses.

**Powerful Hope**

Powerful hope motivates us, as agents, to push through adversity and work to solve our critical problems. Its power comes from a pragmatic vision of the future that really matters to us—especially one that invigorates us with moral passion—and from imagined pathways to that future that we find exciting. The vision must reflect clearly defined values, goals, and identities that bring humanity together around a compelling common purpose, including broadly construed principles that have always been the bedrock of healthy and humane societies: opportunity, safety, justice, and a common feeling of "we-ness."

COMMANDING HOPE'S THREE COMPONENTS are woven tightly together. To be astute, our hope must first be honest: if it's to incorporate a deep understanding of how diverse people might be motivated to address the future's dangers, our hope must realistically acknowledge those dangers in the first place. And to be powerful, our hope must first be astute: if it's to be centered on a vision of the future that brings humanity together, and if it's to catalyze people to search for strategies to make that vision a reality, our hope must reflect a deep understanding of people's worldviews and motivations.

### WHY IT MATTERS

After years of examining the complex problems of societies in crisis, and sometimes in collapse, I'm convinced that commanding hope is the kind of hope we need now in an increasingly disorienting and frightening time—and that our children will sorely need tomorrow.

The double entendre implicit in "commanding hope" is intentional: the term is both a directive to act and a description of the nature of hope we require. It highlights two arguments that bind together everything you'll encounter in the coming pages: First, while we usually assume that "hope" is a general label for an everyday, emotionally pleasant, but largely static state in our minds that we can't control or change, hope is, in fact, something very much under our direction, our command. Both the basic nature of our hope and whether we possess that hope are products of our agency, through which we can assert our intelligence and imaginations. And second, the kind of hope I advocate here is meant to be compelling, so it commands our attention and aspirations—and those of others too.

Ultimately, "hope" is an idea we can reimagine and reinvent. And by turning it into a positive and powerful tool for change that's compatible with diverse cultures, worldviews, and walks of life and that helps us live together more humanely in this century, we can help ourselves and our children flourish in a world in turmoil, and maybe even tame that turmoil to create a far better future.

This isn't the "pessimism of the intelligence, optimism of the will" that early-twentieth-century Italian Marxist Antonio Gramsci famously advocated.[21] (Gramsci adopted the phrase from the great

French writer Romain Rolland.) Commanding hope doesn't seek to temper intelligence's pessimism with the will's optimism; instead it anchors both in resolute realism.

Yet it also keeps this realism from becoming resignation. It intervenes between our honest understanding of the gravity of the threats we face and throwing up our hands in defeat. We can know that the situation is daunting, that we may suffer grievously, and that any good outcomes are profoundly uncertain, but we will also recognize that there's something much larger at stake, something more beautiful, more dignified, and more necessary that resignation would only repudiate.

In the process, commanding hope can give us both the hot motivation and the cold resolve—combined, a kind of ardent tenacity—to keep looking for answers to the world's converging problems, answers that sustain our benevolence and wonder and our belief in the possibility of real human progress. It can give people and societies the moral passion and common purpose they need to reverse our slide towards calamity. Such a hope can also keep at bay the despair that might otherwise suffocate our imaginations. For if imagination creates the mental possibilities that act like oxygen for hope, it's also true that hope creates psychological space in which our imaginations can thrive.

# 6

# Imagine Possibility

*Things look bleak. The propensity to despair is strong, but should not be indulged. Sing yourself up. Imagine a world in which you might thrive, for which there is no evidence. And then fight for it.* Gary Younge

"I CAN'T BELIEVE THIS!"

"What's that, Ben?"

I was making dinner in our kitchen, which opens onto our family room, where Ben was sitting on a couch. He was reading a children's book about the world's oceans—*First Encyclopedia of Seas & Oceans*—that we'd bought when visiting an aquarium.

"Up to seventy million sharks are caught each year," he read carefully. "Their fins are used to make soup. The rest is thrown away."

It was late October 2011. Ben was six and already reading well. But I was pretty sure he didn't know how big a number seventy million is—except that it's really, really big. I certainly didn't expect what happened next.

Coming into the kitchen, he stood right in front of me, arms straight and stiff by his sides, hands clenched into two little fists. "This just isn't right," he exclaimed, red in the face. "I can't believe it. We have to do something." And then, after a few consoling words from me that clearly didn't help, he dashed upstairs to his room.

I had food on the stove, so I didn't follow. Twenty minutes later he reappeared holding a drawing he'd just made of a submarine. The vehicle had a transparent pilot's dome on top, viewing station at the front, propeller at back, and fins on the sides. It also had an articulated arm ending in a nasty-looking claw. "We'll use this claw," Ben declared, "to cut the fishing lines of the people catching the sharks."

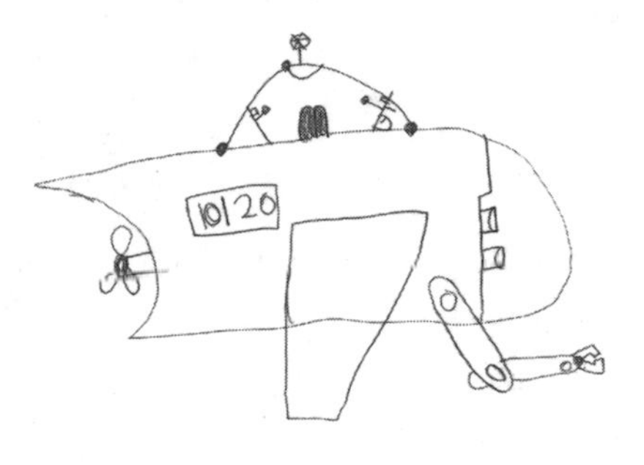


Now, in case anyone thinks Sarah and I brainwash our kids with radical environmentalism, let me be clear: although I suppose we could be classified as North American middle-class environmentalists, we do try to encourage a diversity of views in our family about the best relationship between humans and nature. And at the time of this conversation, we hadn't talked to our children much about Earth's woes. For Ben and Kate, nature was still a source of wonder, mystery, fantasy, and magic. The North Pole wasn't melting—it remained an icy place where Santa lived. We didn't want our children to feel sad, so like many parents, I suspect, we postponed telling them about environmental crises as long as we could.

If Ben's reaction to the news about sharks wasn't inspired by any explicit prompting from us, that made it all the more fascinating. Where did it come from?

On reflection, I decided it sprang from his emerging moral impulses. Ben hates seeing animals and people suffer, and he also has a strong sense of fairness. The news that people kill a huge number of sharks each year just for their fins threatened his world in a personal and immediate way. Perhaps he even made the connection in his mind that if people can hurt sharks for such a silly reason, then they can probably do similar things to other people, even to him.

Then, with his picture and story about the submarine, Ben turned himself from a passive observer of something horrible into an active agent that could stop that outcome. He used his imagination to tell a story that brought the circumstances under psychological control and made them far less threatening. And sure enough, after he drew the picture, he was much less exercised. He went to work with his Lego, and we didn't hear about sharks for the rest of the evening.

That wasn't the end of the matter, though. Within a few days, he'd organized his entire grade-one class to write letters protesting overfishing around the world, particularly of sharks. At first Sarah and I were just surprised observers, but eventually we helped to send all the letters to Canada's minister of fisheries and to the fisheries branch of the United Nations Food and Agriculture Organization in Rome.

A few years later, I asked Ben what he wanted to be when he grows up. It was the first time I'd asked the question directly. "Oceanographer," was his matter-of-fact reply. His response on that earlier occasion—drawing a picture of a line-snipping submarine—had evolved into something much bigger in his mind. To use the language of the late cultural anthropologist Ernest Becker, he'd imagined for himself a hero story that gave his life purpose and meaning.

It's easy to scoff at our children's ambitions, because they often sound silly. But then again, maybe our kids—by imagining hero stories that give them at once hope and a sense of purpose to help make the world better—are doing something many of us have lately forgotten how to do.

#### CROSSING EDGES

Ben's submarine story was a "How about" declaration of possibility. Arising in his imagination, it served as an object—a vision of a desirable world—that became a motivating goal for his hope TO change the future.

Ben's reaction also showed how hope operates along and across the edges of our reality. Hope is a liminal phenomenon. The ancient Greeks seem to have understood its essential "edginess"—Elpis was, after all, trapped at the lip of Pandora's jar. Today, I think this property of edginess is a key source of hope's ambiguity and contentiousness, but also potentially of its power.

I've become fascinated by edges. They include, of course, the physical edges in our daily lives—like those between a road and its shoulder, between the inside and outside of our homes, and between what we can see and what we can't, because it's around a corner. But it's life's metaphysical edges that really intrigue me, like those between what we know, more or less, and what we don't really know at all; between the past, present, and future; between events inside our minds and outside; and between the impossible and the inevitable.

All four of these metaphysical edges play key roles in our hope, and I could see them in Ben's response. Most obviously, his story focused his hope both beyond the edge of the known into the unknown and beyond the edge of the present into the future. He also used his story, and the hope it engendered in him, to bridge his inside-outside edge—to reconcile his internal sympathy for sharks with the external reality of their slaughter.

And finally, the future state he imagined—one with a little submarine busily cutting fishing lines—fell in the boundary zone between the impossible and the inevitable, at least as he perceived it. Scholars have long known that if we're to hope for something, we need to believe its probability lies somewhere between these two extremes. The thirteenth-century Christian philosopher Thomas Aquinas observed in his *Summa Theologica* that "the object of hope is a future good, difficult but possible to obtain."[1] Psychological research supports the observation: a study of university students' subjective experience of hope concluded that "the objects of hope tend to fall in the middle range of probabilities."[2] This is the zone of uncertainty.

We're deeply ambivalent about the first of these edges—the edge between the known and the unknown. It simultaneously captivates and scares us, because beyond it lies a domain of emerging novelty and unexpected combinations—a place that the complexity scientist Stuart Kauffman has wonderfully called the "adjacent possible."[3] Hope is one way we turn this unknown terrain into something less scary: just as Ben did, we use our imaginations to leap across the edge into the unknown future and highlight there a possible world we desire, thus bringing it back across the edge into the known.

But for our hope to be honest, that possible world must be more than just desirable: it must reflect our best judgment about how things could turn out. And if we're going to reimagine and reinvigorate hope in ways that help us, we must think carefully about the relationships between time, imagination, possibility, and prediction.

### TIME TRAVEL

To be able to cross the edge from the present into the future, we need to first go to the past; and to make use of what we find in the domain of the unknown, we need to first start with the known. And in both cases, our vehicle for these trips is an imagination tempered by the knowledge and clear-eyed realism that honest hope demands.

When we recall something that happened to us in the past, as when I think of Ben describing his submarine, we're using what cognitive psychologists call "episodic memory." It consists of images, sensations, and emotions associated with specific events we've experienced, as distinct from "semantic memory" of concepts and facts. With episodic memory, our brain's hippocampus links information from three other brain regions to create an integrated recollection of what happened, when, and where. It's an impressive feat, but if the memory in question remains purely objective—as if we're separate from the remembered event and viewing it from outside—it's ultimately little different from a video recording with a time-and-place stamp.

Our brain performs true magic, though, when it makes the memory subjective. Then, it seems like we're inside the event and part of it. For an instant we *feel* the way we felt at that moment; we have a fleeting sense that we're living it again, as if we've been instantaneously transported backwards in time. When I recall Ben showing me his picture, he's standing in front of me again in the kitchen, holding the sheet of paper, and I feel the same mix of perplexity, wonder, and pride I felt at that very moment. On such occasions, our brain is doing much more than simply viewing a streaming image that has a time-and-place stamp. It's engaged in what scientists call "recursive thinking," something that's almost certainly a hallmark of human cognition—unique to us as a species.

Recursive thinking involves "recursion," and recursion isn't an easy concept to grasp. Computer scientists use it to describe what happens when a computer program invokes or "calls" itself. In psychology, recursion is most easily understood as thinking about thinking. And although this cognitive feat is a bit mystifying, we perform it all the time: not only do we regularly think about our own thinking, we also think about our thinking about other people thinking, and we think about other people thinking about *our* thinking—one can continue ad infinitum in loops of increasing complexity.

Recursive thinking lets us travel in time. On those occasions when we feel like we're living a past moment again, we're conscious of our consciousness at that moment. As the psychologist Michael Corballis says: "In remembering episodes from the past . . . we essentially insert sequences of past consciousness into present consciousness."[4] And what we can do for our memories of the past, we can also do for our visions of the future: we can insert sequences of our imagined future consciousness into our present consciousness.

But in this case our mental feat is even more amazing, because what we imagine hasn't occurred yet, and maybe never will. Our imagination takes chunks of episodic memory—what we recall from the past—separates these chunks from their original contexts, and then recombines them in new configurations. It uses these chunks almost like letters of an alphabet to create the sentences of stories we generate about possible worlds—stories in which we're sometimes central protagonists, or subjects. These stories then become vehicles to explore the future and what it might mean for us, which can help us decide what to do in the present. As the late French biologist and Nobel Prize winner François Jacob has written: "Our imagination displays before us the ever-changing picture of the possible. It is with this picture that we incessantly confront what we fear and what we hope."[5]

Our imagination does face some practical limits, of course: the further its stories depart from our everyday knowledge, memories, and experiences, the hazier these stories generally become in our minds and the harder they are to share with others, which can reduce their utility as starting points for conversations to explore the future. Writers, painters, filmmakers, and other artists trying to depict the future have always struggled with these limits. George Orwell's classic novel *1984* and Stanley Kubrick's movie *2001: A Space Odyssey* both had to use—as part of their storytelling alphabets—common conceptions of technology at the time, which is one reason these works seem oddly quaint now, albeit still immensely powerful and astonishingly imaginative. A series of French artworks published at the turn of the twentieth century titled *En L'An 2000* (two shown here) are also reminders of how our ability to imagine the world tomorrow is shaped by what we know in the present.

*An aerial battle*

*At school*

Any time we use the phrase "hope to," we're taking ourselves on a similar mental voyage to the future. Even when I say something as mundane as "I hope to plant my vegetable garden tomorrow," I'm combining bits of memory of my past experiences gardening and then using the linked set of memories as a vehicle to project my consciousness into the future. For an instant in my mind, I'm in my garden tomorrow, boots on, trowel in hand, and digging in the dirt.

### PUSHING BACK IMPOSSIBILITY

But here's where things get tricky. We can hope for something in the future only if we're uncertain about whether it'll come to pass, as we've seen. If we know for sure the outcome *won't* happen, then hoping for it doesn't make sense. If, on the other hand, we know for sure the outcome *will* happen, then we won't hope for it, because hoping isn't necessary.[6] The uncertainty in the zone between the impossible and the inevitable creates a mental space; our imagination can then populate that space with desirable possibilities, some of which we can make objects of our hope.

Alas, good prediction—prediction that accurately estimates future outcomes—is antithetical to uncertainty, because the better our predictions, the lower our uncertainty about the future. And honest hope, if it's truly honest in the way I've described, must be informed by our best predictions. As our predictions of the future improve—because our scientific theories capture reality's inner workings more accurately, and because we have more and better data about the world—the space for imagined possibilities that can be objects of our *honest* hope narrows, especially when we learn that some desirable outcomes we thought were possible are almost certainly impossible. Although Ben could imagine his submarine cutting the fishing lines, and I might imagine planting my vegetables, those futures are off limits to our honest hope if we have strong reasons to believe they can't happen. (Of course, Ben, being only six at the time, couldn't have been expected to know his "How about" wasn't possible on the mass scale he imagined.)

This is the underlying explanation of today's tension between honesty and hope. The stubborn facts about our global reality—about things like worsening climate change, economic inequality, mass migration, political extremism, and authoritarianism—are making many futures we desire unlikely, even impossible, according to our best predictions.

Once again, it's as if the space for hope is being squeezed ever smaller as the walls of our problems close in—the way I felt that evening when I was overlooking the sea. So, many of us, desperate to maintain some of this space, increasingly ignore the stubborn facts and predictions—those closing walls—and tell ourselves the situation isn't so bad. This behavior has become pathological and self-destructive. It's one factor behind the spread of misinformation about, for instance, supposedly widespread corruption among climate scientists, endemic laziness in poor communities ("welfare queens"), and criminality among immigrants. These "fake facts" allow people to blame a specific group for the worsening problem in question, which helps them maintain their hope that, if only the group responsible would change its ways—or if only it could be controlled, excluded, or even eliminated—then the problem would go away.

But there are honest, fair, and constructive things we can do to keep open the space for hope. For example, as we saw in the last chapter, we can remind ourselves that in a world of complex systems we often don't know enough to be sure that a desirable future, like the peaceful end of apartheid, is impossible. More assertively, we can use our imaginations, tempered by the best knowledge and predictions available, to try to judge more accurately the edge between what we really can and can't change. Sometimes outcomes that we've accepted as impossible in reality aren't, because we can still do something, as agents, to make them happen. In these cases, any verdict of impossibility is premature.

If our hope is to be honest, we shouldn't ignore the facts or deny or avoid well-grounded predictions that a future we might want is impossible—say, a future where climate warming is stopped at 2 degrees Celsius. But we shouldn't just passively accept these predictions either. Instead, we need to vigorously interrogate them and to push back against impossibility. With imaginations still well-informed by the best evidence and science, we can explore the feasibility of worlds in which we've changed the apparently unchangeable. We can examine the future's possibilities objectively—at arm's length, as one might say.

Stephanie May didn't accept that nuclear testing was inevitable, unlike many people at the time; nor did she accept that the superpower conflict was cast in stone. Instead she used her imagination to explore how things could be different, and how she and like-minded activists could make them different. In her memoirs, she wrote:

> The fallout drifted thousands of miles—all around the globe—and fell without respect to national boundaries. What the world needed, I thought, was an international lobby of women who would protest against the fallout which man (or men) internationally were creating.
>
> But how could such a lobby be formed? To be successful it would have to have the most influential women in the world behind it. I envisioned such a lobby. It would be called The Lobby for Humanity. . . .
>
> In October I started writing letters to influential women throughout the world—to Mrs. [Eleanor] Roosevelt, Lady

> Winston Churchill, to the women representing both Chinas—Madam Chang Kai Shek and Madam Sun Yat Sen (for the lobby would have to put humanity before politics), to the wife of the President of France and in India to Madam Pandit.[7]

This first campaign didn't succeed. But the responses she received encouraged her to persist in imagining how to make possible what many thought impossible.

Today, we face a similar situation. It's now widely assumed by social and political commentators, for instance, that the deepening ideological polarization in Western societies between groups on the political right and left is a permanent and inescapable feature of our world's social landscape. But I'll show how, with the aid of new social science knowledge about the ways people's worldviews work—knowledge that my colleagues and I used to develop our tools for mapping worldviews—we can use our imaginations to discover new and feasible ways to bridge the gulfs between these groups.

All of this may sound a bit like the often-quoted Serenity Prayer that the American theologian Reinhold Niebuhr wrote in the middle of the twentieth century.

> God, grant me the serenity to accept the things I cannot change,
> The courage to change the things I can,
> And wisdom to know the difference.

I'm suggesting something more radical and, I think, far more liberating. Niebuhr believed we need an external agent such as God to help us see clearly the boundary between what we can change and what we can't. And he implied that this boundary is fixed, that there's nothing we can do about it, so we can only serenely accept it. With both assertions, he downplays our agency.

And while wisdom, whether granted by God or not, might help us see what we can and cannot change, it's not remotely enough by itself. Wisdom is a rich intuition distilled from accumulated past experiences and knowledge; it's contemplative and often passive. Imagination, in contrast, is a continually unfolding mental experiment that's oriented towards the future, towards opening up possibility; it's engaged and active—and an implicit assertion of our agency.

While wisdom can provide imagination with valuable letters for its alphabet, in the form of recognized patterns drawn from experience and a sense of balanced judgment of what's possible, it's imagination that uses that alphabet to tell stories that can be the basis for our honest hope. Wisdom and imagination play equally vital roles helping us discern what's open to positive transformation.

### CREATING "HOW ABOUTS"

In addition to pushing back against premature verdicts of impossibility (so we can see where we might change the apparently unchangeable), we can also, as noted earlier, look for and even create entirely new possibilities for our future—"how abouts" that we've not previously imagined, like Ben's submarine.

Once again, the complex nature of our world comes to our aid here, because complexity expands the number and variety of novel, unexpected combinations in the adjacent possible, the zone of the unknown beyond the edge of the known.

It's easiest to understand how by looking at our ability to anticipate the way a complex system will change through time. We all know intuitively that when we're dealing with a system like our town or city—social systems that are most certainly complex—this ability degrades quickly the further we cast our mind into the future. Our mental image of future events becomes murkier and murkier. When we imagine events near the present—say, tomorrow's traffic patterns in our city or next week's municipal council debate—we can reasonably assume they'll resemble equivalent events today or in the recent past. In other words, we can hold many of our city's parts and processes constant in our minds and then focus on the few aspects we think might change.

But as we cast our imagination ever-further into the future—to a year or even to ten or twenty years from now—we should acknowledge that many things we've taken as "constants" will change, increasingly so the further we voyage into imagined time. Our city's traffic patterns will shift as new buildings, roads, signals, transit systems, bike lanes, and crosswalks are added; as some two-way streets become one-way, or vice versa; and as new technologies, like electric scooters and autonomous vehicles, appear. Also, ten or twenty years from now, some of the issues before our municipal councils will be radically different—in the 1990s, no municipalities were discussing whether to declare a climate emergency; now towns and cities around the world are doing so.

The further we voyage into imagined time, too, the more we need to acknowledge that factors we've not anticipated at all—like the COVID-19 pandemic that hardly anyone foresaw in 2019—will play steadily larger roles. As the overall number of changing elements expands into the future, the possibility for new, unexpected combinations among the city's multitude of social, institutional, and technological components grows even faster. Add to this mix the inevitable nonlinearity of complex systems—the fact that occasionally small changes will produce huge effects, as when a council member's illness tips the vote on a critical municipal issue—and the range of potential paths our city could follow into the future widens explosively the further we look forward.

We have trouble even envisioning many of these paths, especially decades hence, let alone accurately judging which of them our city is most likely to follow. Some experts in risk and probability call this psychological relationship to the future "deep uncertainty." It seems to have become acute in recent decades, as our world has become exponentially more complex, densely connected, and hyper-kinetic, and as our demographic, environmental, economic, and social stresses have multiplied.[8]

When faced with deep uncertainty, it's human nature to reach for what we know. And what we know—or at least what we *think* we know—is the direction we're currently heading. So, we tend to extrapolate from our recent path into the imagined future, suggesting, essentially, that the future will be a more or less straightforward extension of that path. For instance, if our city's population has been growing at a regular pace for many years, we can project that rate of change forward and then imagine what the city will be like with the predicted rise in the number of people in ten, twenty, or even fifty years. Well-run societies use such projections to make plans for new school construction, water and sewer infrastructure, training of doctors and nurses, and the like.

This way of predicting the future is really just another way of holding things constant, because we're holding a trend constant. It often produces useful results, and I'll employ it myself in the next chapters, because many kinds of change do occur in the same way and at the same rate, or at a reasonably predictable change in rate, for long periods. Yet by definition, it can't anticipate any big and sudden shifts in a complex system's behavior—for instance, the flips from one state or equilibrium to another I mentioned before. And our world today has lots of this kind of change.

In the realm of technology, think of the sudden arrival of the World Wide Web, which was nonexistent in 1992 and widespread by 1998, and which in just six years fundamentally altered the way hundreds of millions of human beings communicated. In the realms of politics and economics, think of the collapse of the Soviet Bloc in the early 1990s, the global financial crisis in 2008 and 2009, the rise of populist nationalism across the West more recently, or the economic upheaval caused by the novel coronavirus in 2020. Far more often than in the past, it seems, momentous surprises—sometimes broadly beneficial—can now burst out of the blue, most clearly in humanity's social affairs. Currently, even our most astute observers are almost always terrible at predicting these sudden changes—even if in retrospect the precursors are often clear.

So, the complex nature of our world can itself create new and unexpected possibilities for our hope. These possibilities may be inherently hard to see before they arrive, but if we can focus our resources on better understanding how the complex systems around us work, we may be able to see some of these possibilities in advance and actively work to realize them. As I'll discuss later, we may even exploit the enormous leverage inherent in today's hyper-nonlinear world to shift humanity's path in a radically more positive direction.

An important additional feature of our complex *social* world multiplies this potential leverage. Most things in our social world—traffic patterns, municipal councils, courts, parliaments, stock markets, and international organizations like the United Nations—exist only because enough people believe they exist and then act in light of those beliefs. In contrast, most things in the physical or natural world, like mountains, oceans, and air, exist regardless of whether human beings believe they exist. For example, our countries are real only because we share roughly the same beliefs about their reality.[9] If someone magically eliminated from all human minds all beliefs about the United States, the United States as a country would no longer exist. Of course, the physical things that we currently think of as being parts of the United States—its landscape, buildings, and people—would still be there, but the country itself wouldn't be.

We can take advantage of this feature by experimenting in our imaginations with radically new beliefs about our social organizations, group identities, political institutions, laws, and norms. And if we communicate some of these new beliefs among ourselves and adopt them widely enough, we'll literally create a new social, economic, and cultural reality for ourselves. If, for example, we can imagine and communicate widely around the world new, powerful forms of the basic idea that our shared identity as human beings overrides our narrower national, ethnic, religious, and class identities—social "facts" or concepts whose importance many of us accept unquestionably today—then that new global identity—that new kind of humanity-encompassing "we"—will start to become part of our dominant social reality, and our collective behavior towards people, societies, and cultures has the potential to profoundly change.

In many ways for the better, I contend.

## THE OPEN FUTURE

There's a final feature of our reality that can help us create new possibilities for honest hope—new "how abouts" that we've not previously imagined—and in some ways it's the most fundamental feature of all: time.

Time is real. That statement probably seems almost silly to most of us. Of course time is real: we live in time and experience time every day, so it must be real! That's just common sense.

Yet among physicists, who think a lot about time, the overwhelming consensus has been that what we perceive as time—chiefly time's flow from the past through the present into the future—isn't in fact real, in the sense that it's not a fundamental feature of the universe. Rather, it's an illusion: our perceptions of time's flow and of the distinction between past, present, and future are artifacts of the way our brains construct our experience of reality.

Physicists have taken this view largely because they're committed to two underlying assumptions about the nature of the universe. The first is that reality is governed by natural laws, like Newton's laws of motion, the laws of thermodynamics, and the equations in Einstein's theory of relativity. And the second is that these laws have stayed the same and held true everywhere in the universe from the first moments of its existence at the Big Bang and that they'll continue to hold true forever.[10]

If the laws that rule reality are universal and eternal, then they've not only determined all past events that happened in the universe, they also preordain all possible future events. In principle, with complete knowledge of these laws and the present state of the universe, we should be able to derive or "compute" all past and all future events. Most modern physicists adopt essentially this idea: that we live in a "block universe" in which all events that have ever occurred or ever could occur in the entire universe—for all the past and all the future—exist together in the same space at the same instant.[11]

The idea of universal and eternal natural law is immensely influential in our modern world. Almost all scientists—not just physicists—accept it at least implicitly. One could even say that the broad acceptance of this idea is one of modernity's defining characteristics. It's also perhaps modernity's misfortune that this idea leads to the conclusion that all future possibilities are already determined, that the future is closed to true novelty and that, more perniciously, our latitude to exercise our agency is tightly bounded. "A world without time," writes the brilliant physicist Lee Smolin in his book *Time Reborn*, "is a world with a fixed set of possibilities that cannot be transcended. If, on the other hand, time is real, and everything is subject to it, then there is no fixed set of possibilities and no obstacle to the invention of genuinely novel ideas and solutions to problems."[12]

A generous and thoughtful polymath in his mid-sixties, and an authority on quantum gravity, Lee Smolin was one of the founders of the renowned Perimeter Institute for Theoretical Physics in Waterloo. I count it as one my life's great blessings to have met him when I moved to the Waterloo community. Lee is among those rare scientists both willing and able to convey the subtleties of science to the public, and he has written several popular books explaining the inner workings of modern physics. He's also something of an iconoclast, challenging the orthodoxy of his corner of the scientific community, most famously with his sharp critique of string theory, which is still physics' dominant explanation of nature's fundamental forces.

But his most sweeping and subversive challenge to conventional physics is his argument that physicists need to rethink time—its deep nature and how it's represented in their discipline's theories and research projects. He argues persuasively for the reality of time in our lives and universe, and as a new foundation for theoretical physics, and he extends the argument's implications far beyond physics into all the other sciences and into moral and political questions in our everyday lives. (He developed this argument in collaboration with Harvard University's legal and political theorist Roberto Mangabeira Unger; together, they've written an academic book, *The Singular Universe and the Reality of Time*.)

For the sciences, perhaps the most significant implication is that if time exists fundamentally, then natural laws that are universal and eternal probably don't exist fundamentally.[13] Their argument has enormous implications, too, for our thinking about hope: if time isn't real, then to keep our hope honest we must focus on ensuring that the objects of our hope—the future states we desire—are among those that already exist in the block universe. Honest hope can only involve imagining as accurately as possible a future that's already present. So it will tend to be passive—a hope THAT a specific future already exists and will come to pass.

But if time is real, then the future doesn't yet exist, which means it's open to true novelty. The act of hoping can also involve *creating* new building blocks of reality itself, as we first imagine and then bring to fruition genuinely new possibilities. So our hope can indeed be active—we can hope TO create, as agents, a desirable future.

The late German philosopher Reinhart Koselleck held that the idea of an open future emerged in the West only in the late-eighteenth century.[14] By the mid-twentieth century, the relationship between this idea and scientists' concept of time had become a matter of lively debate. In conversations with Albert Einstein, the philosopher Karl Popper argued for the reality of time. "I tried to present . . . as strongly as I could my conviction that a clear stand must be made against any idealistic view of time," he writes in his autobiography. "And I also tried to show that, though the idealistic view was compatible with both determinism and indeterminism, a clear stand should be made in favor of an 'open' universe—one in which the future was in no sense contained in the past or the present, even though they do impose severe restrictions on it."[15] Another thinker who argued for an open future was Oxford political theorist Isaiah Berlin, who in the 1950s famously launched a full-throated attack on historical determinism as both cognitively incomprehensible and morally indefensible.[16] But I find the Smolin-Unger thesis more convincing than Berlin's: by reconceptualizing time, it attacks the heart of the argument used by physics, today's dominant scientific discipline, to deny the possibility of an open future.

This isn't mere semantics or a metaphysical digression, like so much debate about the nature of time through history. Lee Smolin, for instance, is very clear on the political and social implications. He eloquently notes that a civilization "whose scientists and philosophers teach that time is an illusion and the future is fixed is unlikely to summon the imaginative power to invent the communion of political organizations, technology, and natural processes—a communion essential if we are to thrive sustainably beyond this century."[17]

### "YOU DON'T HAVE THE RIGHT TO TAKE AWAY MY HOPE!"

So exclaimed a young woman to her mother, an American friend of mine who's an expert on climate and energy issues. My friend had just described to her daughter how hard it will be to change our energy systems fast enough to prevent global warming's devastating effects. Sometime later, my friend mentioned to me her daughter's response. That was years ago, yet the words have kept coming back to my mind.

That's partly because I empathized with her daughter: hope is a crucial psychological lifeline for most of us, which means we're understandably afraid and angry when other people say things that might cut that lifeline. But her words have come back, too, because they raise uncomfortable questions: one set of questions we might ask ourselves in a situation like the daughter's, and another set in a situation like my friend's.

If we find ourselves in the daughter's situation, what right or entitlement do we have to hope when truth threatens our hope? More pointedly, what's our responsibility to listen to and accept the truth?

And if we're in my friend's situation, what exactly is our responsibility to sustain people's hope when we have accurate knowledge about dangers they face? And if we don't have a right to take away their hope, are we then obliged to actively hide the truth from them, or perhaps deceive them in some way, so they can hold on to some kind of hope—even if we know it's false?

At the time of the conversation, the daughter was already an adult. Most of us probably believe, at least at first thought, that a competent adult should be able to hear the truth, even if it threatens some of their hopes, and particularly if the person providing that information has expertise on the topic, as well as some responsibility for the other's well-being. So, for example, we probably think that a patient should be willing to hear accurate information from his or her doctor about a serious illness like cancer. My friend is an expert on climate change, a problem that threatens her daughter's future well-being, and my friend clearly has some responsibility for her daughter's well-being. For these reasons, my first reaction was that my friend was right to tell her daughter what she knew, and that her daughter shouldn't have resisted hearing that information because of its implications for her hope.

But then I thought some more, because questions like these are now acute for researchers, such as climate scientists, who are experts on the potentially cataclysmic dangers humanity faces. In fact, they're at the center of a controversy that has split the environmental movement.[18] On one side are people, including many scientists in the relevant fields, who believe that the truth should be told, even if it's dismal. On the other are those who think that dismal truths simply scare people and cause them to retreat into helplessness. These folks say that if we want people to do something about a dire problem like climate change, then it's usually better to soft-pedal its seriousness, even if doing so means hiding or bending the truth a bit.

But neither side seems to fully recognize that two critical factors underlie this controversy. The first is one's degree of certainty that the outcome will be extremely harmful, should the problem continue unaddressed. Even if we know a problem like climate change *might* cause us great harm, we cannot be completely certain that this harm will happen if it arises from a complex system. A doctor might have only a rough sense of the probability that a kind of cancer will kill a patient. Is the doctor then being dishonest and irresponsible if he or she emphasizes that uncertainty to sustain the patient's hope? Surely not always. When my fellow writer and I were asked by that audience member years ago "Is it too late? Is it hopeless now?", I answered that exploiting uncertainty's fuzziness about the future in order to sustain hope can be an honest thing to do.

But at what point does it become dishonest?

The second factor underlying the controversy over how much truth should be told is one's estimate of the difference one could make by trying to address the problem. Of course, lots of uncertainty often surrounds this factor too. But usually we can come up with a reasonable if rough estimate of the difference.[19]

By splitting the two factors into two values—which is, admittedly, an arbitrary thing to do—we can set up a two-by-two table, like the one shown here, which gives us a sense of how the factors might combine:

- If we want to sustain hope and our motivation for concerted action, cell 1 in the table is the worst place for us. There, we're simultaneously certain that the future outcome will be horribly bad for us and that we can't do anything to make it less bad; in this situation, it's truly impossible for us to hold in mind a motivating object for our hope—a clear vision of a desirable future.

- Cells 2 and 3 are intermediate positions, where hope is threatened but might nonetheless survive.

- Cell 4, in contrast, is the best place for us. There, we believe the outcome might not be so bad and that, in any case, we can do something to make it better.

| | We believe we can make no difference. | We believe we can make at least some difference. |
|---|---|---|
| We're certain that the outcome will be extremely harmful. | 1. Hope dies here. | 2. |
| We're uncertain that the outcome will be extremely harmful. | 3. | 4. Hope lives here. |

*Factors affecting the life or death of hope*

When we understand complexity—that we usually don't know the full possibilities of change and novelty in the complex systems we live in and that small differences in these systems can dramatically shift the path we take with those systems into the future—we help move ourselves honestly from the table's top row of cells to its bottom row. When we emphasize that the future is open and that with the aid of our imaginations we can see better how to change it or maybe even create it (rather than just discover a future that already exists), perhaps by leveraging our world's complexity, we help move ourselves honestly from the table's left column of cells to its right column. And if we use *both* approaches together, we help shift ourselves, again honestly, towards the bottom right corner—to cell 4—where hope can live.

This line of thinking made me realize that my initial reaction to the daughter's response was too judgmental. Now I think that she had a point, but mistakenly framed it in terms of her exclusive right, rather than in terms of rights and responsibilities on both sides of the conversation.

If we find ourselves in something like my friend's situation, we surely have the right—a natural right, grounded in defensible moral principles—to speak the truth, if we have access to it, regardless of what that truth might be or its implications for others' hope. But with that right, I now realize, comes a responsibility to help people—as best as we can—cope with that truth emotionally by explaining how honest hope might still be possible, perhaps through their taking meaningful action.

If instead we find ourselves in the daughter's situation, we might not have a right to hope itself, but we do have the right to have our *need* for hope respected, based on an expectation of common decency of treatment. But with that right comes the responsibility to listen to truth and to adjust our hope in the context of that truth—to, in other words, keep whatever hope we can find honest. As we'll see later, this can mean radically changing the object of our hope by substituting a new desirable future in the place of the one we had originally.

And if, in the end, we conclude that all honest hope is impossible—that we're inescapably in cell 1—then we have the responsibility to accept that apparent reality and not retreat into false hope.

But we're not there yet.

# 7

# Courage beyond the Edge

*Tell no lies. . . . Mask no difficulties, mistakes, failures. Claim no easy victories.* Amílcar Cabral

PARTICULARLY IN THE WEST, HOPE has a bad rap in popular discourse. As fear, frustration, and anger become ever more dominant social emotions, many people seem to regard conventional ideas of hope with ever more disdain.

Critics fall into three main camps. Some think hope is *vague* and *naive,* because often it doesn't have a clearly defined object—a vision of a desirable future. Such was the gist of Sarah Palin's "hopey-changey" ridicule of President Obama's idea of hope. Others think hope is inescapably *false*, because even when it has an object, it rests on a belief in the mere possibility of that desired future. By this view, the uncertainty that's essential to hope encourages people to fantasize about good outcomes, which then encourages wishful thinking about the likelihood of those outcomes. Patrick Henry, one of the "Founding Fathers" of the United States, purportedly expressed this sentiment in a speech to the Second Virginia Convention in 1775: "It is natural for man to indulge in the illusions of hope. We are apt to shut our eyes against a painful truth and listen to the song of that siren till she transforms us into beasts."[1]

And a final group of critics argues that hope is a dangerously *passive* response to the world's challenges, because by focusing our attention on (at best) merely possible outcomes, hope distracts us, keeps us from acting to change the world, and thus diminishes our agency. The late United States rear admiral Gene La Rocque provided a striking example of this view in an interview in the early 2000s:

> I spent my life planning, training, arming, practicing, and fighting in wars. I spent seven years in the Pentagon trying to find better ways to kill people, destroy things. I was a strategic war planner. I tried to find more ways to kill people all over the world. I spent seven years in war colleges teaching people how to kill people, destroy things. I never once let myself think about hope. There was no hope. I was looking for certitude. . . . If we want a better world, we as human beings ought to do what we can to bring about the change. Hoping is a futile mental exercise.[2]

La Rocque was a remarkable person—a survivor of the Japanese attack on Pearl Harbor and over a dozen battles in the Pacific theater and, despite the cold-blooded statements above, later in his life a tireless advocate of nuclear détente and civilian control over the military-industrial complex. Nevertheless, it's easy to see how, for a military man like him, hope could appear to be a frivolous, even deadly distraction. In the midst of combat, just hoping for a good outcome is apt to get one killed. What one needs to do is rapidly and accurately appraise what's happening, weigh available options, and then act.

La Rocque's kind of skepticism about hope—that it encourages passivity—has deep roots in history. Hope indicates "a lack of knowledge and a weakness of mind," said the Dutch rationalist philosopher Baruch Spinoza, a pioneer of Enlightenment thinking. "The more we endeavor to live by the guidance of reason, the more we endeavor to be independent of hope."[3] And recently, the American author and radical environmentalist Derrick Jensen has written that:

> Hope is what keeps us chained to the system, the conglomerate of people and ideas and ideals that is causing the destruction of the Earth. . . .
>
> The more I understand hope, the more I realize that all along it deserved to be in [Pandora's] box with the plagues, sorrow, and mischief; that it serves the needs of those in power as surely as belief in a distant heaven; that hope is really nothing more than a secular way of keeping us in line.
>
> Hope is, in fact, a curse, a bane. . . . [It] is a longing for a future condition over which you have no agency; it means you are essentially powerless.[4]

So, should we just give up on hope? Some great philosophical traditions, like Stoicism, have argued that hope is a distraction that only leads to unhappiness. The Roman philosopher and playwright Seneca emphasized the close association of hope with fear. "Both," he wrote, "belong to a mind in suspense, to a mind in a state of anxiety through looking into the future. Both are mainly due to projecting our thoughts far ahead of us instead of adapting ourselves to the present."[5]

Today, as our future grows steadily darker, relinquishing hope and focusing on the present appeals to more and more people. The anthropologist Michael Taussig notes that modern intellectuals tend to associate lack of hope with "being profound."[6] Says Jensen: "A wonderful thing happens when you give up on hope, which is that you realize you never needed it in the first place. You realize that giving up on hope didn't kill you. It didn't even make you less effective. In fact it made you more effective, because you ceased relying on someone or something else to solve your problems."[7]

Paul Kingsnorth wanted to help people relinquish hope when he co-founded the Dark Mountain Project—an international network of people trying to make better sense, through shared stories, of the emotional and moral import of humanity's environmental crisis. Everyone wants hope, he notes. "Hope, all the time. Hope, like a drug. Do not look down—look away." He and others began the Dark Mountain Project because "we needed to look down, and not to flinch as we did so."

> This Project was created to build a place, a scene, a space, where people could mass to, among other things, talk openly about what they saw when they looked down and, if necessary, share that sense of despair without feeling that they had to leaven it with talk of hope, campaigning for change, goals, movements or activism.[8]

In sum, Jensen argues that relinquishing hope can enhance our agency, by helping us break with the dominant power system, so we can better challenge that system and get on with the work of actively finding solutions. Kingsnorth stresses that relinquishing hope can allow us to more completely acknowledge our despair. Both Jensen and Kingsnorth say that despair is an entirely reasonable reaction to the crises humanity faces, and I agree. I'd also say that we need to consciously recognize and accept this despair, because it's unhealthy to bury emotional trauma deep in our psyches.

But all these critics are mistaken, I believe, in their disdain for hope, because they're not considering the qualities of what I've been calling commanding hope. First, to keep our hope from being vague and naive, it must have a clear vision of a positive future. Then, to keep it from being false, we must avoid wishful thinking about the likelihood of that future.

Our hope isn't necessarily false because it rests on a belief in the mere possibility of a desired future. It *becomes* false when we ignore or select evidence in order to convince ourselves that those outcomes are more likely than they actually are. And we can choose not to do this: we can have an exciting possibility in our mind but still be ruthlessly realistic about its likelihood, drawing on the best evidence and predictions we have available. False hope doesn't arise from imagining we'll win a lottery; it arises when we convince ourselves—despite evidence to the contrary—that our chances of winning are significantly more than zero and then use that wrong estimate to make winning an object of our hope.

So while I'm sympathetic to the critics' insistence on reason, realism, and honesty about likelihoods, I also think they're profoundly wrong when they imply that we should entertain in our minds only those future outcomes that seem highly likely, given what we see around us—that, to paraphrase La Rocque, we should always seek certitude. If we adopt this approach, we won't use our imagination to explore the broad range of less likely outcomes, some of which could offer us the chance of a much better future. Nor will we imagine, as Mahatma Gandhi, Martin Luther King, Nelson Mandela, and Stephanie May did—and as Greta Thunberg and her fellow activists around the world are doing today—how we might use our agency to make some of those less likely outcomes real. And that ensures they'll stay less likely. This argument holds, I believe, even in harsh, cold circumstances such as those of military decision-making. While at some point, particularly in war, the exploration of alternative outcomes must stop, and final decisions must be made, a general unwillingness to imagine better futures can bring closure too soon.

Our capacity and need for hope, as long as we keep hope honest, is a precious gift, because it encourages us to keep open a space for possibilities, and to use our imagination to create possibilities in that space. That hope thrives on uncertainty is not a weakness but its greatest strength.

And finally, we can keep our hope from being dangerously passive by making sure it's "hope TO" not "hope THAT." When Jensen declares that hope is "a longing for a future condition over which you have no agency," he's talking only about "hope THAT." And when he further says that hope's object is always a vision of the future created and sustained by society's power elites, he omits the possibility that we can hope for an outcome in which the current system and its future manifestations are replaced. There's no reason why our compelling vision of the future—the object and goal of our "hope TO"—can't help us recognize the ties that bind us to the ideas and ideals of dominant elites, give us an alternative future to strive for in our political struggle with those elites, while at the same time guiding how we allocate our limited resources to maximize the chances we'll win.

Rather than disdain hope, as these critics do, we should treasure it. Positive psychologists have shown that few of us can flourish physically and psychologically without hope. Our reasonable and necessary despair mustn't displace our hope, making despair the final stage of our response. That would be psychological capitulation—as if we're kneeling before fate and baring our neck for its sword.

We don't need to give up on hope, either, because we can give it commanding new meaning and make it work for us in new ways. We can avoid despair by honestly acknowledging the critical nature of our situation, while still imagining positive alternatives that work within our future's likely constraints. We can be strategically smart as we strive for those alternatives by becoming astutely informed about other people's perspectives. And we can make that positive future real by articulating a vision of the future that powerfully motivates our agency.

### THROW OURSELVES INTO WHAT'S BECOMING

No one should doubt Gene La Rocque's courage, of course. But he was also wrong to imply hoping is a sign of cowardice—or in Spinoza's words, "a weakness of mind."

On that point, the skeptics have it exactly backwards: courage and hope often go together, as do despair and cowardice. Aristotle keenly observed that "the coward . . . is a despairing sort of person; for he fears everything. The brave man, on the other hand, has the opposite disposition; for confidence is the mark of a hopeful disposition."[9]

In coming years, our hope for a good future will face ever more severe trials. As we experience ferocious droughts and storms, dying ecosystems, immense migrations, and declining economic security and political stability, we'll need to push back against despondency and fatalism, all without lying to ourselves. It will take all our courage to hope honestly and never to give up trying to make things better.

Hope always draws its power from operating along and beyond reality's edges. But to realize that power, we'll need the courage to learn how to hope well—to exercise and strengthen, in a sense, our muscle of hope to succeed in that liminal environment. "The work of this emotion," the German Marxist philosopher Ernst Bloch wrote in the 1950s, "requires people who throw themselves actively into what is becoming, to which they themselves belong."[10]

Bloch was saying, I think, that we'll need courage to travel beyond the present's boundary into the unknown and to use there our recursive imaginations to combine chunks of what we know in an explosion of genuinely novel possibilities. We'll need it to see how, beyond that edge, we might produce the sharp shifts in our societies that will create a more positive future. We'll need it, too, to accept the magnitude of the problems we face, while seeking the right balance between uncertainty and prediction and between imagined possibility and reality—a balance that can keep open a space for honest hope. And we'll need courage to admit our own ignorance while at the same time believing that through our agency, individually and collectively, we might still make a difference.

Most importantly, we'll need courage to hold on to the possibility of a good future for our children and to the conviction that they, too, can have some reasonable hope. If we aren't willing to give up, to say it's all pointless, but we also refuse to escape into ignorance, denial, or magical thinking, then we should get to work with our imaginations to find a way out—to identify desirable yet realistic futures and ways to reach them.

But first, there is a particular arc of imagination we should be cautious about, because it can mislead us and doesn't point to a way out of catastrophe.

# PART TWO

# THE CHALLENGE OF HOPE

# 8

# The False Promise of Techno-Optimism

*Science in the service of humanity is technology, but lack of wisdom may make the service harmful.* Isaac Asimov

I SPENT MUCH OF MY YOUTH hiking along the remote shores and through the wild forests of Vancouver Island—often alone and with little more than a compass and ax. The landscape is in my bones. The sound of north-Pacific surf crashing against cobble beaches and rocky bluffs; the scent of western red cedar, Douglas fir, and the thick, damp soil of the forest floor; the rough texture of moss and lichen hanging from branches like beards; the crunch of arbutus leaves underfoot in late summer—these sensations still call up some of my deepest memories and emotions.

In June of 2016 I took Ben and Kate to a very special place there—an isolated beach on the island's extreme western coast that I'd visited with my parents when I was twelve. That day in the summer of 1968, we'd walked several miles along the shore to a heavily wooded point. My father, a forester with vast experience surviving in the coastal forest, had hacked a path across the point, through undergrowth thick with devil's club, ferns as tall as a person, and salal so dense and high that it blocked out the sun.

The scene that greeted us as we came down from the forest onto the sand was magical. The beach was isolated by massive headlands at each end of a bay and backed by a high cliff topped with impenetrable old-growth rainforest. In the middle of the bay was a small rocky island. Because the tide was low, the beach consisted of two crescents that met at a raised sandbar connecting the shore to the island. The sea in each crescent was crystal blue and quite calm. I could see clouds of spray rising out to sea, where a distant reef sapped the energy of the oncoming Pacific waves.

It was paradise for a child. As the tide rose through the day, the sandbar slipped underwater, and the two beaches merged, so the island was isolated from the shore. The waves crossed over the warm sand as they swept around each side of the island, creating a splendid pattern. I spent hours there jumping and dancing in the tall peaks of water created by the intersecting waves, with sea spray shimmering around me in rainbows.

When I took Ben and Kate to the same spot almost half a century later, the beach looked just as I remembered it. But now it's part of a national park, and we arrived via a long trail through the woods from the nearest road, and down a wooden stairway of hundreds of steps that descends the cliff behind the beach. The tide was low once again; and as it rose, the two beaches became one, the waves crossed, and the kids jumped and danced in the glorious sun and sea, just as I'd done so many years before. They taunted the incoming sea with shouts and song, daring it to come closer, and marveling as it infiltrated and then swept away their sandcastles.

As I sat on the beach and watched, my mind traveled back and forth between the present moment and that day far in the past. Then it flashed into the future, as I imagined how Ben and Kate might bring their own kids (or grandkids) to the beach some fifty years hence to play in the crossing waves—say in 2068, a century after my first visit.

Which abruptly snapped me back to today's reality. I realized that the information available now indicates that the beach will look very different fifty years down the road.

According to today's best scientific estimates, the world's seas will be much higher, as global warming causes water in the planet's oceans to expand and glaciers to melt. The rate at which the ocean's average level is currently rising—about three millimeters a year—seems miniscule. But as years go by these increments add up; and in any case, the rate is going to accelerate sharply. Depending on how the ice sheets on land in Greenland and Antarctica respond to warming and on future emissions of greenhouse gases, the seas could rise a meter or more by 2068, according to recent studies.[1] They'll then continue to rise for many centuries, even if humanity's carbon emissions fall to zero, because it takes hundreds of years for sea level to fully adjust to a warmer atmosphere. Indeed, emissions to date have already increased Earth's surface temperature to that last seen during the Eemian period (115,000 to 130,000 years ago), when sea level was six to nine meters higher than it is today.[2]

Warming is also supercharging the atmosphere and oceans with energy, creating vastly more powerful storms. Together, by later this century higher seas and bigger storms will have scoured many coastlines of the sand and gravel that make their beaches.

Not just the beach will change radically. The old-growth forest behind it, including trees half a millennium old, will decline and die, too, as coastlines erode, summer droughts desiccate the landscape, new warmth-loving pests arrive, and sick and dead trees burn in wildfires. (This process has now begun: just two years after our visit to the beach, in the summers of 2018 and 2019, heat and drought killed hundreds of thousands of cedar, fir, and arbutus trees across southern and central Vancouver Island.) All in all, the present's facts tell me that just fifty years from now my magical beach and its surroundings will be almost unrecognizable.

And as I sat watching Ben and Kate, I had to admit it had already changed, but in subtle ways. When I'd visited in 1968, the little rocky island in the bay had been festooned with sea stars—small and large; brown, red, orange, and purple—amidst masses of mussels and sea anemones all clinging to the wet rock revealed by the low tide. With their kaleidoscopic colors, bizarre shapes, and creeping movement, the animals were an integral part of the place's magic. But in June 2016, we could find only four small ones hidden in a rocky crevice at one end of the beach.

Beginning around 2013, sea star populations collapsed from southern California to Alaska, succumbing to a syndrome eventually labeled "sea star wasting disease." Scientists aren't exactly sure of the cause, but new research strongly suggests that warmer seas have made the creatures more susceptible to a viral epidemic.[3] Maybe the sea stars will eventually come back—and by 2019 a few clusters were reappearing in scattered locations along the coast—but their near-disappearance is nonetheless a striking sign of deep change in the coastal ecology of western North America.

AT THE END OF OUR beautiful day, as we brushed the sand off our feet, put our shoes back on, and trekked up the steps, I couldn't stop thinking about the fate of Ben and Kate's generation. How can we help our children develop a realistic sense of their future? How can we help them strike the right balance between uncertainty and possibility on one side, and facts, reality, and some degree of solid prediction on the other? How, in other words, can we best ensure that the vitality of their recursive imagination—and of ours—is tempered by the clear-eyed view of the future that honest hope demands?

One way we can temper our imagination is to identify key features of humanity's situation that we're pretty sure will persist in the foreseeable future.

These "constants," as I called them earlier, create a structure of opportunities and constraints for humanity—just the way the structure of a road system creates both opportunities for our travel, by making it easy to go to some places, and constraints on that travel too, by limiting access to places without roads.[4] But how do we identify the relative constants in humanity's situation today?

We could examine forecasts people have made in the past about a future that's already happened, or is happening now—using documents or stories that have come down to us—and look for patterns in how those forecasts have turned out to be right or wrong.[5] Yet it's actually hard to recall accurately how people once thought about the future, because assumptions about what's plausible in the future change as time passes, often subtly: what once might have seemed a likely development, such as the advent of flying cars, comes to look silly; while other outcomes that might have seemed outlandish, such as carrying a fabulously powerful computer in a pocket, start to appear as if they were always inevitable. This is what psychologists call a "shifting-baseline" problem.

In our own lives, it helps to have experienced a sharp, punctuating episode that's still alive in our memory and associated with recollections of how we imagined the future at that time. As it turns out, a tragedy in my childhood gives me such an opportunity today.

### THE FUTURE, IMAGINED IN THE PAST

A little over eighteen months after my parents and I, aged twelve, visited that magical beach, my mother was gone. Beginning in the spring of 1969, a massive attack of multiple sclerosis ravaged her body. Her sight started to fail, and within two months she couldn't walk. A wildlife illustrator, she found the loss of fine-motor control of her hands devastating.

It was an appalling, harrowing time for all of us. But it had beautiful moments too. My mother was a biologist by training, as well as a writer, sculptor, and photographer, and she was well-read across dozens of fields. She kept her faculties till near the end, and because I knew she was desperately ill and declining fast, I spent as much time with her as I could. We talked about everything under the sun—art, literature, poetry, science, politics, and history. On July 20, 1969, I sat on the end of her bed, as she, my father, and I watched—on a small portable black-and-white television—Neil Armstrong step onto the moon.

We talked about the future then, and I can recall distinctly how we saw it. Today, I often imagine what things would—and would not—surprise my mother about the state of our world, if she were to visit five decades later.

I know she'd be surprised, first, by how little most of the basic technologies in our lives have changed. Sure, we now have devices like microwave ovens, push-button phones, and flat widescreen color televisions. But with one important cluster of exceptions, our everyday technologies are largely the same as they were half a century ago, just better and sometimes bigger—or much smaller. They're basically modifications of earlier technologies. My mother would immediately recognize the refrigerators, stoves, plumbing and lighting systems in middle-class houses; the cars in the garage (and the internal-combustion engines that still power most of those cars); and the expanded urban and industrial infrastructures, energy grids, and transportation networks that we've laid down on the planet's surface over the last fifty years.

The exceptions would be the information technologies that have become inextricable parts of our lives, like desktop computers, smartphones, and tablets; the internet, web, and social media; and search engines, GPS-guided maps, and street-view apps. These technologies are radically new compared to 1960s cassette recorders and analogue land-line telephones. Here we've seen sharply non-linear, revolutionary change. My mother, with her appetite for knowledge and a lifelong desire to travel, would initially be captivated by devices that can instantly show us virtually any fact or place in the world. And having done all her writing on a Remington manual typewriter, I know she'd gape in wonder at the magic of word-processing and speech-to-text software.

But I suspect that after she'd played with these technologies for a few days, disappointment would start to replace wonder. After all, Kubrick's movie *2001* had been released in 1968. We went to see the movie together, and she talked me through its more abstract scenes. The movie reflected the widespread conviction in the West that technological progress was accelerating. Science-fiction writers, computer scientists, and the public alike were absolutely sure, for instance, that computers would be able to talk and think like people within a decade—two decades at the outside—just the way the mainframe computer, HAL 9000, does in the film. My mom wouldn't be surprised or impressed by Siri, Alexa, and Google Assistant today.

And speaking of space, she'd be shocked—after watching the moon landing in 1969—that no human being has gone beyond low-Earth orbit in nearly fifty years. We were supposed to be on Mars and maybe even as far as Jupiter or Saturn by now. But a half-century has passed, and our principal sources of energy remain hydrocarbons; solar and wind generate a relatively small percentage of our energy; fusion power remains fifty years in the future; cancer hasn't been beaten; we still can't predict earthquakes; and despite all the high-tech drugs and gadgetry of modern medicine, life expectancy in wealthy countries is only modestly greater—nine years for men and seven years for women in the United States; among some US populations it's even declining.

And multiple sclerosis is still wrecking people's lives.

In the technological realm, then, my mother expected spectacular change, so I know she'd be very surprised today by the relative absence overall of such change. In the political, economic, and social realms, though, I imagine her reaction would be exactly reversed: she expected continuity for the most part, and she'd be amazed by the degree of change.

The Vietnam War was at its peak—the North Vietnamese and Viet Cong launched the Tet offensive in early 1968—so she'd wonder what happened to state communism, which dominated a large fraction of the human population at the time and seemed a formidable and durable political force. The collapse of the Soviet empire and the end of the Cold War would shock her too, because in the 1960s these likewise seemed to be fixtures of the international system. And she'd be fascinated by the rise of China as a mighty capitalist power and the stunning jump in its population's wealth; in the late 1960s, China had been in the middle of the Cultural Revolution and seemed perpetually mired in poverty and ideological turmoil. Then she'd learn about the collapse of apartheid in South Africa, the expansion of the European Union, the huge increase in global trade and finance, the Cambodian and Rwandan genocides, the soaring participation of women in the workforce, the widespread acceptance of gay marriage in Western countries, the growing recognition of gender diversity, the rise of suicide bombings, and, of course, the elections of both Barack Obama and Donald Trump as presidents of the United States.

Other aspects of today's political, economic, and social realms would surprise her less, including the growth in the human population (she'd notice immediately the crush of people and traffic in her city), the profound divisions and hatreds among religious and ethnic groups in many regions, the continuing bitter poverty of hundreds of millions, the weakness of international institutions, and the corrupt governments, failing economies, and wars that are driving millions of people from their homelands. She'd see in these things continuity from the 1960s.

But the thing about today's world that would surprise her the least—while breaking her heart the most—is the vast ruin humanity has unleashed on its natural environment.

My mother wasn't a soft-hearted tree-hugger: among her talents, she was a crack shot and a superb hunter. Yet she marveled at and treasured nature. As a child on Florida's beaches in the 1930s, she collected thousands of seashells. I still have them, wrapped in tissue paper and stored in old shoeboxes; some are now exceptionally rare, because the species of which they are representatives have largely vanished there. I also have hundreds of slide photographs she took of Vancouver Island's indigenous wildflowers. As with the shells she collected, many of the wildflower species she photographed are now hard to find or gone completely.

Even though in the late sixties the public's awareness of environmental issues was limited—the environmental movement was just beginning, and the first Earth Day wouldn't happen till 1970—my mother sensed the looming threats. She knew about Rachel Carson's book *Silent Spring* and the asphyxiating smog in big American cities. In the year before she got sick, she helped me with school projects on air and water pollution, taking me to a local plywood mill and a sewage outfall to document and photograph effluent flows. She grasped that humanity was hurting the planet, and she talked to me about it at length. Although the problem of climate change was little known at the time—only in 1965 had scientists delivered to US president Lyndon B. Johnson the first high-level report on the matter—she'd immediately understand the science today. And its implications—for the natural world, for her grandchildren, and for all humanity—would make her weep.

I think we can draw from this exercise three key insights about persistent features of our situation that will affect our future possibilities. First, despite all the everyday media hype about the blinding speed of technological change in the modern world, since the middle of the last century almost all truly revolutionary advances have occurred in information technologies like smartphones, the internet, and search engines, while changes in other domains, from transportation, construction, and medicine to agriculture have been largely incremental, and sometimes halting at best.

Sweeping waves of revolutionary advances—across most if not all technological domains—have happened in the past, most notably from the mid-nineteenth century to the 1970s, when an unprecedented series of radically new and useful technologies disrupted advanced economies: the internal combustion engine and the automobile, rockets, airplanes, electric lighting, telephones, radio, television, refrigeration, air conditioning, indoor plumbing, urban water-treatment and sanitation, antibiotics, hybrid grains, petrochemicals, grid-distributed electricity, and nuclear power—the list goes on. My mother was born into the middle of this revolution, which is why she and just about everyone else in the 1960s didn't harbor a scintilla of doubt about the ferocious momentum of technological progress. But waves of innovation across multiple technological domains turn out to be exceedingly rare; indeed, nothing like the one I've just described had happened before in human history.[6] Far more common were extended periods when technological change involved step-by-step improvements in already existing basic technologies—which is exactly what we've seen since the 1960s in most areas except information technology.

Second, contrary to conventional wisdom, change generally happens far more rapidly in the social realm. It's in our social systems—not in our technologies—where we more often see the dramatic, nonlinear shifts that are truly revolutionary in scope and implication. Empires collapse, financial markets crash, political revolutions erupt, mega-corporations go bankrupt, and even people's fundamental values can change dramatically, sometimes with breathtaking results—as we've seen recently with the rapid shift in some societies' attitudes and laws regarding equal rights, gay marriage and, much less happily, with surging populist nationalism around the planet.

And the third insight: compared with fifty years ago, when palpable damage to the natural world seemed almost always local or regional, with just the first rumblings of a truly global environmental crisis, our species has increasingly become a dominant force altering chemical flows, landscapes, and webs of life around Earth. The environmental changes we're causing are now so immense and widespread that they're operating at the planetary level, are visible just about everywhere one looks, and are starting to rebound on us in critical ways, in the form of terrifying wildfires, enormous storms, dying coral reefs, declining pollinator populations, and countless other environmental dangers.

**ASSUAGING OUR ANXIETIES**

These insights hint at both the opportunities and constraints that will shape humanity's future. As we try to imagine possibilities that can sustain our hope in that future, we can use the lifespans of today's children as a reasonable time scale. The year 2068, as I realized on the beach, is the same distance into the future as the late 1960s are in the past, and a point when children like Ben and Kate will be in their sixties. So, what will their lives be like then?

The world could be a wonderful place. In 2068, Ben and Kate's generation could be secure, healthy, and happy, living in free and just societies, with satisfying jobs, full-grown kids of their own, and lots of time and money for travel, leisure, learning, and entertainment. Some popular thinkers today—Erik Brynjolfsson, Ruth DeFries, Peter Diamandis, Gregg Easterbrook, Andrew McAfee, Ray Kurzweil, Matt Ridley, and Steven Pinker among them, nearly all (it's worth noting) male and from the West, and many associated with Silicon Valley—think there's every reason to believe that today's children can enjoy this kind of future, largely regardless of where they live in the world.[7] I call these thinkers "techno-optimists," because they generally highlight scientific and technological solutions to humanity's problems. Although they admit that a few clouds lurk on our horizon, they're full of hope—of a kind—as they look towards coming decades, because they're quite confident that the forces driving human progress are sufficiently powerful to sustain that progress indefinitely.

Some, such as Steven Pinker, are more careful with reasoning and evidence than others, such as Gregg Easterbrook.[8] And while their individual arguments differ in important respects, they commonly assert that because human reason and creativity know few limits, we can justifiably expect our lives to keep getting better. The spread of free markets and democracy and the advent of scientific breakthroughs in biotechnology, artificial intelligence, robotics, nanotechnology, quantum computing, brain-machine interfaces, nuclear energy, and the like can generate near-universal prosperity and help solve global problems like climate change, poverty, and war.

Here's a typical statement, in this case by Peter Diamandis and Steven Kotler from their 2012 book *Abundance: The Future Is Better Than You Think*:

> Humanity is now entering a period of radical transformation in which technology has the potential to significantly raise the basic standards of living for every man, woman, and child on the planet. Within a generation, we will be able to provide goods and services, once reserved for the wealthy few, to any and all who need them. Or desire them. Abundance for all is actually within our grasp.[9]

We probably all wish this vision of the future could be realized, for our children's sake. And if we squint a bit when we look at our species' history, we can see the evidence to justify this hope. Human beings are astoundingly good at inventing technologies to better their lives, even if most invention through the millennia has been incremental. And that extraordinary wave of truly revolutionary technologies that arrived from the mid-nineteenth through the mid-twentieth centuries has indeed delivered to billions of people around the world, and continues to make possible, huge improvements in basic material well-being, like enough calories each day and much lower risk of deadly disease.

That technological wave didn't occur in isolation, though, as Steven Pinker particularly emphasizes. It partly arose from, and its beneficial impact was multiplied by, widespread improvements in social arrangements, institutions, and practices, like the rule of law, increased literacy, public education and health care, more responsive and competent states, and a liberal economic order. Also, and less positively, the rapid progress of some key technologies like rockets, airplanes, and nuclear power was propelled by major wars, both hot and cold. Still, the overall result has been a surge in general welfare, including far greater personal income, freedom, and dignity for much of humankind.

According to the World Bank, the proportion of the world's population living in extreme poverty (defined as daily consumption equivalent to $1.90 or less in 2011 prices), for example, fell by nearly three-quarters, from 36 percent to 10 percent, between 1990 and 2015, which is likely the fastest decline in history.[10] Nearly half the world's population now qualifies as middle class, with enough money to provide for basic needs and at least a little left over for luxuries. These gains help explain why the world—through the last half of the twentieth century and until about 2010—saw a steep fall in the global incidence of death (measured in terms of deaths per one hundred thousand people per year) from armed conflict such as interstate and civil wars, insurgencies, ethnic clashes, and genocides.[11]

To see how dramatically conditions have changed—at least in some parts of the world—just walk through an old cemetery in a wealthy Western society. In the one behind my house in Ontario, many of the headstones dated before 1900 are for children under five years old, and a heartbreaking number are for babies less than a year old, because children often died from illnesses we've since learned to treat. Many of the headstones from 1914 through the early 1950s commemorate young men killed in war. But after 1960, all these types of death almost vanish from the headstone record.

Yet, if we look more closely at the evidence, we find that the techno-optimists' vision of the future is almost always highly selective. It might assuage some of our anxieties about the state of the world, but it's radically incomplete and so, in many respects, wrong. And while it might seem open-minded about future possibilities—in the ways I described in the last chapter—it in fact often lacks exactly the combination of wisdom and courageous imagination we need to challenge deep conventional assumptions and create space for truly novel possibilities.

Most importantly, techno-optimism, in its less reflective forms, can encourage us to be passive in the face of the world's astonishing challenges, because it implicitly suggests that we don't need to be concerned much about the future or do much to make it better, and that we don't need to act on injustices within and among our societies. Sweeping, inexorable, and impersonal forces—technological progress, in particular—will take care of everything for us.

## INFATUATED WITH BRAINPOWER

Techno-optimists generally build their argument on a cluster of assumptions, conclusions, and ways of using data that, I would argue, taken together, make their hope false.

First, they're convinced human beings are exceptional, in the sense that our rationality, creativity, and problem-solving ingenuity allow us largely to transcend the biological and material limits that affect other species. Bluntly, those who hold this view are infatuated with human brainpower, particularly when this brainpower is organized and focused within free economic markets. They point to the remarkable technological revolution from the mid-nineteenth through the twentieth centuries to justify claiming that, given enough time and the right incentives, humanity can supply brainpower in near-endless quantities and should ultimately be able to address all its key challenges—from climate change and resource shortages to economic inequalities—successfully. Our future, in short, is boundless.

This assumption is less wrong than dangerously simplistic and misleading. It leads, for instance, directly to the idea that we're separate from and superior to nature. By this view, our minds can see through and understand the myriad complexities of Earth's variegated natural systems, from fisheries and forests to insect ecologies and the climate itself, including these systems' feedbacks, nonlinearities, and hysteresis (that is, their capacity for irreversible flips to new states, as when a fishery collapses or, as may happen in the future, when Earth's climate reorganizes itself in a way that's catastrophic for agriculture). Then, it's only one small step to the conclusion that we're smart enough to manage, manipulate, or replace at will—for our exclusive benefit—all natural systems, as if they were everyday machines.

Infatuation with human brainpower and reasoning also leads techno-optimists, I believe, to overestimate the rate and scope of future technological change. They take a recent trend—the astonishing progress in information technologies, for instance—and extrapolate it into the future; then they assume that the same degree of future progress will happen across all technological domains, from biotech and materials science to nanotechnology. For these folks, recent advances in information technology offer incontrovertible proof that the far wider technological revolution over the last two centuries hasn't ended, and in fact has never slowed. It was just the first stage of an indefinitely long era of explosive scientific development.

By my reading of the history of technology's progress, this is wishful thinking. Sure, we'll see some remarkable changes in coming decades. For instance, after years of false starts, artificial intelligence is finally advancing quickly and, in the process, making a range of other technologies feasible, like fully autonomous military weapons and civilian vehicles. We'll likely see major strides in energy and energy-conversion technologies, too, as the effects of climate change bite. Already the cost of some forms of renewable energy, like solar photovoltaic electricity, has fallen far faster than experts predicted even a few years ago; new kinds of nuclear power, such as breeder reactors that generate their own fuel internally, are emerging; and the electrification of vehicle fleets is well underway, despite the relentless opposition of fossil-fuel interests.

Yet in many domains where the techno-optimists anticipate breakthroughs, we'll probably see something like what happened over the last fifteen years in the field of human genomics. When the genome was fully mapped in the early 2000s, scientists and commentators were thrilled, because they thought researchers would soon identify the genetic causes of many awful illnesses, like cancer and Alzheimer's disease, and then finally develop true cures. Yet this early promise hasn't panned out: as researchers have worked to understand the genome, they've learned that the genetic origins of most illnesses are far more complex, subtle, and contingent than they'd first thought.[12]

So the rate of technological change this century is likely to be patchy—sometimes rapid, but slow and steady in most domains most of the time. And we should be skeptical about bold claims that a sweeping wave of revolutionary advances—across most if not all technological domains—is just over the horizon. If we were magically transported to 2068, we'd undoubtedly recognize most technologies around us, just as my mother, if transported from fifty years ago, would easily recognize most technologies around us today.

But above all, many of humanity's critical problems can't be solved through simple technological fixes, because they're not at root technological.

Problems like climate change and worsening economic inequality ultimately arise from factors such as the beliefs and values we hold, the design and function of our economies and political systems, our societies' structures of political and economic power, and the obstructionism of vested interests. Addressing these problems requires, at least in part, new beliefs, values, institutions, and behaviors. And to the extent new technologies are part of the answer—in the way that renewable energy from solar and wind power is part of the answer to climate change—techno-optimists often downplay the political, social, and economic roadblocks to implementing them; for instance, our habituation to a consumer lifestyle or the hostility of powerful fossil-fuel companies to the rollout of renewable power.

Finally, the techno-optimists often use evidence—usually statistical data—in ways that can obscure vital but ugly truths about our societies and the human condition. Steven Pinker, for instance, cites statistics like homicide rates or wealth per capita in whole countries or even the entire world to identify general trends in the human condition in the past, present, or future. Such ways of tracking change in humanity's well-being can be revealing—indeed, I used the World Bank's figures on global poverty just a couple of pages ago—but they can also hide gaping disparities in the real-life experiences, conditions, and opportunities of diverse people, groups, and societies around the world.

When techno-optimists say that people's lives have steadily improved—and will likely do so in the future—they frequently neglect the essential questions "Whose lives have improved?" and "Whose lives will improve?" We shouldn't base generalizations about human progress mainly on evidence like that from the cemetery behind my house in Ontario, because the cemeteries in other parts of the world—in African American communities in the US South, in malaria-prone parts of Asia, in civil-war wracked Yemen, or in regions of Central America tormented by narco-violence—look vastly different. And they tell vastly different stories about changes in the human condition through history.

To be fair, to think effectively about the world around us, we do sometimes need and want to refer to the experiences, characteristics, or behavior of large groups of people taken together, even of all humanity. But such generalizations always come with a caveat: we miss crucial details about individual people and smaller groups inside the larger groups, and we miss the fine-grained stories of people's lives.

# 9

# The World to Come Today

*Comprehension [means] examining and bearing consciously the burden which our century has placed on us—neither denying its existence nor submitting meekly to its weight.* Hannah Arendt

OUR SPECIES' DEFINING CHARACTERISTIC may be—as the techno-optimists assert—an unparalleled ability to use brainpower to solve problems, and this ability has undoubtedly been a blessing in uncountable ways. But it's also sometimes something of a curse. Understanding why is critical to seeing more clearly our future's constraints and opportunities.

To an extent, of course, what counts as a problem or solution is in the eye of the beholder; it's affected by that person's circumstances and worldview. A cotton farmer's immediate problem might be boll weevils, and she might think that applying pesticide to her crop is the easiest, cheapest solution to the infestation; while an urban twenty-something might be more concerned about toxic chemicals in the environment, and he might believe that buying natural, pesticide-free fabrics is the best response. In the same way, an environmentalist in a rich city might think that climate change is the world's most urgent problem and advocate for fuel taxes as the best means to cut fossil-fuel use; while a resident of an urban shantytown might be worried about unemployment and believe it's important to keep kerosene, gasoline, and diesel fuel cheap and abundant to sustain businesses and create jobs.

Still, it's worth remembering that whether we're rich or poor our biological and psychological similarities to each other as human beings greatly outnumber and outweigh our dissimilarities. We have in common basic existential needs, and this means that many of our problems are more or less universal across our species. Every day we must all get enough food and water, for example. We all also need shelter from the elements, we want to avoid disease and injury, and almost all of us crave love from our families, approval from our communities, and meaning in our lives.

Decades ago, the American psychologist Abraham Maslow famously examined these similarities.[1] He identified a hierarchy of human needs, with needs directly related to our physical survival at the hierarchy's base, followed by our needs for safety, love, belonging, esteem, and the like. Some regard Maslow's theory as outdated, especially his concept of self-actualization, which is the motivating need at the top of the hierarchy; and we must all recognize that huge numbers of people around the planet have little prospect of reaching the pyramid's top levels, because they're struggling on a day-to-day basis to satisfy their most basic needs. All the same, Maslow's theory has seen something of a revival of late, and his hierarchy remains a serviceable list of what nearly all human beings feel are their key problems and those problems' importance relative to each other.[2]

Across the millennia, human beings have usually addressed problems arising at the base of the pyramid first. When those most fundamental needs aren't met, the problems are easy for people to sense or "see"; they're pressing in time, and nearby in space. If we're

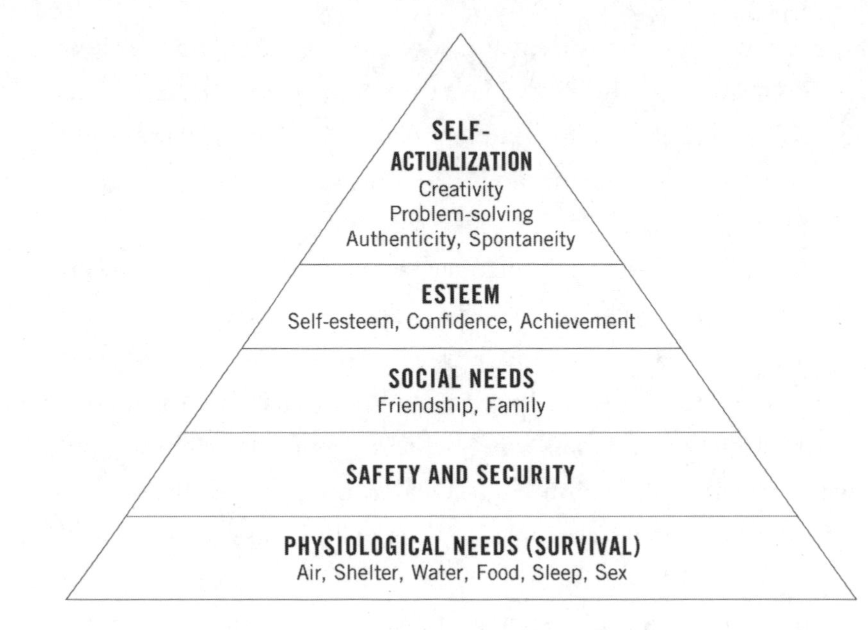


*Maslow's hierarchy of needs*

hungry, for example, we feel it right in our stomachs. So, over thousands of years, people have invested huge amounts of ingenuity to solve the hunger problem, and as a species we've learned how to produce food in massive abundance—even if that abundance is unequally shared.

In the same way, as a species we've figured out how to prevent or cure many infectious diseases, provide ourselves with clean water, house ourselves securely, communicate with each other instantly across long distances, and transport ourselves and our stuff quickly and cheaply nearly anywhere on the planet. And because we hate the vulnerabilities that come with poverty, we've collectively learned how to use economic markets—and the technologies those markets have encouraged us to invent—to generate previously unimaginable material wealth, though once again that wealth is distributed woefully unevenly.

Such problem-solving has allowed us to massively expand our population—about fifty-fold in the last five hundred years—and grow our world gross domestic product (GDP) nearly 150-fold during the same period. For all intents and purposes, we've behaved like a voracious superorganism, spreading across the planet's surface and multiplying our numbers, activities, and artifacts to fill almost all the useful space in the biosphere—that layer of life on Earth's surface as thin, proportionately, as an apple's skin.[3]

Throughout this expansion, as people and groups have struggled to make their lives better, they've understandably paid far less attention to problems that were harder to see, weren't as pressing, or produced impacts far away from them. In fact, we've often effectively swapped the first kind of problem for the second kind: in trying to fix a visible, pressing, and nearby problem, we've helped create others that are much less visible (even invisible), slower to develop (and so less pressing), and more geographically remote. In a sense, we've pushed our problems beyond our horizon of awareness. Of course, our ignorance has often contributed to this outcome as much as our intentions: when we don't understand the downstream results of efforts to fix our immediate problems, those results are in effect invisible to us and so much easier to ignore. But whether intentional or not, the outcome has been the same throughout our species' history: we've consistently displaced our problems beyond what I call our "awareness horizon."

Here's a simple example. Just over a hundred years ago inventors and entrepreneurs, mainly in Western societies, addressed people's visible and pressing transportation problems by developing and mass-producing the gasoline engine. It was a breakthrough: in cars people could get where they wanted to go in a flash compared to trundling across the landscape with a horse and cart, in a horse-drawn carriage, or in a train. But the cars' engines burned petroleum, which produced carbon dioxide, which then helped create a far less visible and pressing problem (at the time)—atmospheric warming. To take another example, to address immediate communication and travel problems, humanity has invented and deployed communication technologies like the internet and cell phones and transportation technologies like jet aircraft and containerized shipping. These advances have connected people, businesses, and economies across larger and larger distances, but they've also helped create "winner take all" markets spanning the whole globe, in turn worsening long-term problems of high wealth concentration and devastating economic inequality.

It's easy to find countless other examples of how people have swapped visible and pressing problems for problems that are less visible and immediate, at least for them. Many farmers feed their livestock antibiotics to promote growth and control disease, but in doing so they exacerbate antibiotic resistance that spawns epidemics of untreatable infections. Medical researchers, health workers, and development agencies have tackled endemic disease in Africa, driving down death rates in societies across the continent—a wonderful accomplishment. But because the less visible problem of the lack of women's empowerment hasn't also been addressed, fertility rates have remained high in many countries, causing rapid population growth that has weakened national economies and contributed to immense migrations to other lands.

Financial and commodity exchanges have linked themselves together with lightning-fast computers and fiber-optic cables to reduce transaction times, but in doing so they've made possible flash crashes that can propagate across exchanges, wiping out monumental wealth in seconds. Manufacturers have installed robots to boost productivity and lower costs (and thus boost their profits), but in doing so they've helped ratchet up chronic unemployment among low-skilled and middle-skilled workers in many economies. And nearly all of us use social media to stay in touch with our friends and keep up on the headlines, but in doing so we've often created, or at least reinforced, hermetic social bubbles that encourage political extremism and undermine the consensus-building processes across groups that democracy needs.

A critical yet often neglected element of this process is social power, which almost always biases humanity's problem-solving. The problems of powerful people and groups are, no surprise, far more likely to be "seen" in the way I've just described—that is, identified as visible and pressing—while the problems of less powerful people and groups, particularly those who aren't regarded as part of "we" by the dominant group, are more likely to be unseen and unaddressed. So the process of swapping visible and pressing problems for less visible and immediate ones is often really a process of making powerful people's problems better (boosting corporate profits with robots, for instance) by making weaker people's problems worse (slashing employment of low- and middle-skilled workers, for instance).

To use economists' language, powerful groups "externalize" many of the costs of solving their problems by shifting these costs to less powerful people, often elsewhere in geographic space or in the future. This in turn results in an enormous transfer of wealth from the weak to the powerful.

### LOCKED IN A DESPERATE RACE

People on the lucky side of these transfers, as the present conditions of our world show, usually simply ignore problems that, at least from their vantage point, are hard to see or less pressing. After all, it's precisely because of those characteristics that the problems are relatively easy to ignore—at least for a while—and with luck, they'll never become bad enough to cause "us" much trouble. But that's just wishful, even magical thinking: ignored problems generally just get worse, and over time they cause all of us, not just those with less power, escalating harm.

For instance, we've known for decades that our carbon dioxide emissions cause global warming but decided to largely ignore the problem; now, that warming is visibly helping to wither crops with record droughts, envelop cities with life-threatening heat waves, inundate low-lying coastlines with high tides and storms, and, of course, smother vast regions in wildfire smoke. We've also known for decades that jet travel would accelerate the global spread of disease but decided to ignore that problem too; now COVID-19 has affected billions of people and devastated national economies.

In other words, our chickens are coming home to roost. Problems we once dumped beyond our awareness horizon—outside our spheres of activity and concern—and then forgot about are now dramatically rebounding on us, with impacts visible to all, even the rich and powerful among us.

It's as if our species has inadvertently locked itself into a desperate race—a race between the increasing number and difficulty of our common problems, on the one hand, and our collective ability to supply solutions that work for everyone or even most of us, on the other.

In *The Ingenuity Gap*, I described a solution as a recipe (or more accurately, as a set of instructions or an algorithm designed to address a given problem). Just as cooking recipes tell us what ingredients to take off our shelves and how to combine them to produce something good to eat, our "solution recipes" tell us what things to take from our natural, technological, and social environments and how to combine them in a sequence of steps to make our problems go away—or at least stop them from getting too bad. Sometimes we might use a specific recipe once, because then we've solved the problem for good; other times, we might need to use a recipe repeatedly. Generally, though, as our problems get harder, we must invent longer recipes to solve them; and as our problems happen faster and in more complex combinations, we have to invent recipes at a higher rate.

Despite what the techno-optimists say, our technological brainpower is only one thing determining whether we'll win this race between our need for good solutions and our ability to supply them. And brainpower won't be remotely enough on its own. Indeed, we're now losing the race badly: evidence from numerous scientific bodies shows that we aren't coming close to adequately addressing many of our most critical problems. We're losing partly because those problems—in their sheer magnitude, complexity, and simultaneity—are growing steadily more challenging with each passing year. We're losing, too, because, as we'll see in the next chapters, several factors are keeping us from supplying solutions to these problems when and where we need them, and some of these factors are themselves becoming stronger over time.

When critical problems such as the climate crisis don't get solved, they can become increasingly powerful stresses on our societies. I call them "tectonic" stresses, because they're like the inexorable forces that accumulate when Earth's crustal plates collide deep underground. And just as these geological forces can erupt into city-shattering earthquakes, so can today's social tectonic stresses erupt into society-shattering earthquakes, such as political revolutions, pandemics, food crises, financial panics, and wars.[4]

## SLOW PROCESSES

Sometimes when I look at my children, I feel I'm looking through a window deep into the future. Of course, given how uncertainty propagates the further we look ahead, I can't see a lot with confidence. Occasionally I glimpse a pattern or a shape, but then that hint of order often vanishes in a swirling fog.

But as I try to imagine what the world could be like in 2068, roughly a half-century from now—or even what it could be just ten or fifteen years from now in 2030 or 2035, given the urgency of some of the recent scientific reports on Earth's environmental crisis—I find it helps to distinguish between slow and fast social processes.[5] The tectonic stresses operating deep in our societies are slow processes, while the earthquakes or system flips that these stresses sometimes cause—those punctuating episodes of sudden social change that experts have so much trouble predicting, like the collapse of the Soviet Union, the 9/11 attacks, or the novel coronavirus pandemic—are fast processes.

We notice the flash and bang of fast processes, but we're generally oblivious to the subliminal creep of slow processes. Still, by continuing over long periods and by contributing to episodes of sharp nonlinear change, slow processes have far-reaching effects.

Today, if we focus on tectonic stresses arising from hard-to-see and less immediate problems that we've let fester, we can confidently pinpoint several slow processes that create a structure of both constraints and opportunities around us. They're the most clearly identifiable, persistent features of humanity's situation (or constants, in the sense of "constant trends") that will shape our future's possibilities.

### Tectonic Stresses in Biophysical Systems

One group of these slow processes operates mainly within the biophysical systems we're part of and have created—as organisms using matter and energy, producing waste, and reproducing within our natural environments. The most important of these is *human population growth.*

Nearly 11 billion people will likely live on Earth by 2100—nearly 50 percent more than the 7.7 billion people here as of 2019.[6] Most of that growth will happen in the next thirty years; and by 2068, other things being equal, the world's population will already be well over 10 billion. A large fraction of the growth will happen in sub-Saharan Africa, with the region's current population of about 1.1 billion projected to double by 2050 and more than triple to 3.8 billion by 2100. Probably about half this growth through 2050 is essentially locked in, even if African women dramatically lower the number of children each has on average in coming decades. That's because large numbers of young girls alive today must still pass through their reproductive years, and even if they have fewer children on average during those years (that is, their fertility rate falls), the total number of births in Africa will continue to rise for decades. This is what experts call "demographic momentum," and it's an extremely powerful (and underappreciated) social force.

Recently, some commentators have pointed to falling fertility rates in many parts of the world to assert that we aren't facing a problem of population growth but instead one of population decline.[7] While it's true that United Nations demographers estimate that populations in some fifty countries will fall by 2050, demographic momentum ensures that, in contrast, the populations in most countries in Africa and West and South Asia will continue to grow for many decades. The result will be a dramatic shift in the balance of the world's total population towards these regions and away from East Asia, especially China.

The size and growth of the human population matter to our well-being together on Earth, but so does what the people and groups in that population do, specifically what and how much they consume in material resources—from oil and iron ore to meat—and produce in waste, including sewage, toxic pollution, and, of course, carbon dioxide. This consumption and waste output are staggeringly unequal across and within our societies. The average Tanzanian, for example, emits one-twentieth the carbon dioxide of the average Swede, one-seventieth that of the average Canadian, American, or Australian, and less than one-hundredth that of the average resident of Bahrain.[8]

While these differences are crucially important from social and moral perspectives, it's our human population's combined impact—the total of all humanity's consumption and waste—that mainly drives biophysical processes around the planet. As we've expanded our numbers, activities, and artifacts into the biosphere's nooks and crannies, we've drawn down its natural capital of ocean fish stocks, old-growth forests, fresh water, fertile soils, and the like. In the process, we've disrupted, and sometimes just overwhelmed, the biosphere's natural ways of regulating itself, including its processes of recycling water and absorbing, storing, and circulating life-giving elements like carbon and nitrogen.

And in the course of these activities we've transferred spectacular amounts of "natural wealth" from the weak to the powerful—in this case from people today and in the future who'll no longer have access to forests, soils, and fish to those of us who, in the recent past and today, have enjoyed, and destroyed, that abundance.

From this macro perspective, the second slow biophysical process that's key is *atmospheric warming* driven by the buildup of greenhouse gases. With emissions so far, we've already warmed the atmosphere at Earth's surface about 1 degree Celsius over its 1850 to 1900 "pre-industrial" average.[9] The stated intention of the 2015 Paris Climate Agreement is to keep this warming below 2 degrees Celsius, with an aspirational ceiling of 1.5 degrees. But by mid 2020, we were already close to slamming into that lower ceiling: it takes a long time for past and current emissions to have their full warming effect, so warming of another several tenths of a degree above our current one degree is already in the pipeline, so to speak.[10] On top of this, even if optimistic scenarios of the rollout of renewable energy are realized, the global economy isn't going to kick its fossil-fuel addiction overnight, which means humanity will probably still be emitting substantial amounts of carbon dioxide well past mid-century. So by 2068, warming of at least 2 degrees is virtually assured—indeed, we're likely to pass that threshold by 2050—and we could easily be on course for 3 or 4 degrees Celsius by the end of the century without a radical shift in our sources of energy and ways of life.[11]

In fifty years, this warming will likely have produced not only a meter or more of sea-level rise, caused the death and burning of many of Earth's forests, and generated a much higher frequency and intensity of extreme weather events—from heat waves and drought to cyclones and floods—it will have also killed nearly all, if not all, of Earth's coral reefs, which support a large fraction of the world's tropical fisheries. Many of these effects are appearing now and could be devastating for innumerable communities in as little as a decade or two.

And in fifty years, some heavily populated regions—including much of the Middle East, the Persian Gulf, and tropical zones in Asia and Africa—will start to experience for part of each year combinations of outdoor temperature and humidity approaching the natural threshold of human survival, a "wet-bulb" temperature of 35 degrees Celsius, or 95 degrees Fahrenheit.[12] A wet-bulb temperature is that shown by a thermometer with its bulb covered by a wet cloth, to maximize cooling from evaporation; a 35-degree wet-bulb temperature is equivalent to 35 degrees at 100 percent humidity. When temperature exceeds this limit, the human body can't expel its own metabolic heat—through sweating and other mechanisms—fast enough to cool itself, and even fit people exposed to such conditions die within a few hours. The stark reality is that regions experiencing such high temperatures will become increasingly uninhabitable—which means that hundreds of millions of people in some of the planet's most densely inhabited zones will have to move to survive.

But what we really need to keep in mind is the impact of multiple stresses happening together. Well before 2068, warming and rising human population will have combined with widespread overuse and pollution of fresh water supplies in the planet's rivers, lakes, and aquifers to cause severe water scarcity for a large fraction of humanity. Warming, water scarcity, and extreme weather (especially heat waves and droughts) will together also threaten world food production, most importantly farmers' output of grains like rice, corn, wheat, and soybeans that provide much of humankind's calories and protein. In a major 2019 study, the Intergovernmental Panel on Climate Change reported that "the stability of food supply is projected to decrease as the magnitude and frequency of extreme weather events that disrupt food chains increases."[13] As the leading crop-science research consortium, the Consultative Group on International Agricultural Research frankly notes: "Securing and maintaining necessary levels of calories, protein and nutrients for populations around the world will be an exceptional challenge."[14]

From our day-to-day perspective, atmospheric warming seems very slow—after all, what does it mean to us if the average surface temperature rises about a quarter of a degree per decade? But it's actually happening faster than almost any change in Earth's average temperature that human civilization has experienced before. Most critically, it's happening far faster than the rate at which our societies' infrastructures and institutions normally evolve. And that's important, because given enough time, we could probably build wealthy, stable societies with a climate that's 2 or 3 degrees warmer than today's. But our problem this century will be the sheer rate at which we're going to have to transition our societies—and the ten-billion-plus people squeezed into them—to the new reality.

And it's not just the human species. Most animals and plants all around Earth are already having enormous trouble adapting to this warming. Their populations will shrink drastically, and many species will go extinct, contributing to the third slow biophysical process of vital importance this century: the broad and steep *decline of biodiversity*, in what is now known as the Sixth Mass Extinction. (The five previous mass extinctions, each of which saw three-quarters or more of Earth's species vanish, took place at intervals between about 450 and 66 million years ago. The causes ranged from huge volcanic eruptions that sharply boosted atmospheric carbon dioxide levels to, in the most recent extinction event, an asteroid impact that eliminated all non-avian dinosaurs.)

Indeed, Earth's web of life is now suffering a staggering barrage of threats on top of climate change.[15] For those of us who marvel at the beauty and mystery of living systems, the result is nothing short of the annihilation of nature.[16] Logging, mining, land-clearing for farms, roads, and housing, and rapacious fishing practices (including driftnet, dragnet, and blast fishing with explosives) are destroying vast tracts of habitat on land and in the sea; higher carbon concentrations in the atmosphere are acidifying Earth's oceans, damaging fisheries and plankton; invasive species spread by globalization are wiping out local plants and animals; and toxic chemicals and persistent pollutants (like microplastics and pesticide endocrine disruptors) are interfering with organisms' feeding and reproduction. Already, scientists estimate, today's extinction rates are one hundred to one thousand times higher than pre-human levels across diverse environments around the planet.[17] Populations of vertebrate species—not just the ones we think of immediately, like elephants, orangutans, and polar bears, but largely overlooked species of hedgehogs and partridges—have declined 60 percent on average between 1970 and 2014; while in the last fifty years in North America, the population of birds, including common species like sparrows and blackbirds, has dropped by three billion.[18] Some of the species that we're losing in whole or in part are vitally important to our economies; pollinators, for example, provide an essential service for about a third of our food production.

The fourth slow biophysical process of note is the *rising "energy cost"* of finding crucial natural resources and making them useful for our economies. People and societies tend to exploit first the most accessible and highest-quality reserves of any resource—whether iron ore, ocean fisheries, or good soil—and then move to more distant, lower-quality reserves. As we use up the best reserves, we dig deeper for less-rich iron ore, we go farther across or deeper into the oceans to hunt down less-valuable fish, and we clear forests off land with thinner topsoil to grow our food. This extra work means we must spend ever more energy to produce or exploit a unit of the resource.

The most fundamental expression of this trend is the rising energy cost of getting energy itself.[19] For instance, in the early twentieth century in the United States, most oil wells were drilled in fields that were close to the surface and often naturally pressurized, so they sometimes produced the "gushers" we see in old industry photographs. Such wells could regularly generate more than a hundred times the energy used to drill them in the first place. Today's Canadian oil-sands mines, in stark contrast, have an "energy return on investment" of about six-to-one.

We can offset this kind of falling resource *quality* by using more resource *quantity*—by digging up and processing more iron ore, for instance, as I outlined in chapter 3. But this response damages more of the planet's surface, which further harms biodiversity; and if it requires us to use more carbon-based energy in the process of seeking energy (a terrible irony), it also worsens global warming. Or, we can use new technologies to exploit previously inaccessible reserves—the way drillers have used hydrofracking to extract oil from deep seams of shale underground in, for example, the Dakotas, Texas, and the English countryside. We might even be able to shift to entirely new resources and technologies to satisfy key needs, as we're already shifting from fossil fuels to wind and solar power. But all these responses imply one thing: we're constantly innovating and working harder to compensate for the declining quality of our available resources.[20]

**Tectonic Stresses in Social Systems**

Slow processes are operating simultaneously within our social systems. The most important is *widening economic inequality*—particularly between the world's poorer and middle classes, on one side, and people in the upper few percent of wealth and income, on the other. In 2017, according to the charity Oxfam, 82 percent of growth in wealth worldwide went to 1 percent of the world's population.[21] The smaller the slice of the world's uppermost economic elite one uses for comparison purposes, the worse the inequality looks: employing data from the investment bank Credit Suisse, Oxfam estimated that the wealth of just sixty-one people equaled *all the wealth* of humanity's poorest half, nearly four billion people.

It's clear to anyone paying attention that economic inequality has risen sharply in wealthy countries in the last four decades, as governments have lowered trade barriers, corporations have shed manufacturing jobs, and returns have flowed more to capital than labor and more to brainpower than to muscle-power. In the United States, despite steady (if historically unremarkable) economic growth over the previous decade, inflation-adjusted wages for the median male worker in 2019 were *lower* than in 1979.[22] Meanwhile, between 1978 and 2016, CEO incomes in the biggest companies rose from 30 times that of the average worker to 271 times.[23]

It is true that during this period in some poorer countries, China and India most significantly, fast economic growth rapidly enlarged the middle class and raised average incomes. One result, as the economist Branko Milanovic has shown, has been a narrowing of overall economic inequality *between* poor and rich countries, as poor countries, especially in Asia, have closed the gap with rich countries in average national incomes. But the story has been very different *inside* both poor and rich countries, where economic inequality between wealthy people and everyone else has generally risen steadily.[24] This trend is unlikely to reverse, because the average pace of economic growth in most national economies, including China's, has slackened, and (for reasons I'll discuss later) will probably stay lower indefinitely. In the world's current economic system, if growth isn't continuous and robust, inequality tends to climb.

Rising inequality magnifies the harm caused by the other slow processes, including the biophysical ones like climate change. People on the poorer side of the widening economic gulf are less able to protect themselves from bigger storms and floods by building better shelter or moving to less vulnerable regions. Meantime, the widening gulf erodes the feeling of common fate—the sense that "we're all in this fix together"—which is a hallmark of the human ability to face a crisis successfully and which might encourage richer people to help poorer people cope with warming's dire impacts.

A last key slow social process affecting our world is *rising migration* of people within countries and across borders. It's driven by the combined force of all the above trends, especially high population growth, low economic opportunities, and extreme weather events like droughts that wreck farmer's crops—all factors that escalate poverty; worsen violent crime, ethnic discrimination, and political dissension; weaken governments; and even generate civil war—in turn forcing people to leave their homelands.

"We are now witnessing the highest levels of displacement on record," said the United Nations in 2017, noting that nearly seventy million people worldwide had been "forced from home," including over twenty-five million people formally classified as refugees.[25]

In the future, if economic inequalities worsen further and global warming increases as projected, the migration pressure on Europe from West and South Asia and from Africa, especially, will become extreme, even before the middle of this century. This pressure will be incomparably more powerful and destabilizing than the recent migrations into Europe that provoked a nationalist backlash. Canada and the United States may be geographically buffered by oceans, but they can still expect large influxes from Latin America and overseas—as we are only beginning to see in the miserable, heartbreaking crush of migrants on the Mexican-US border in 2018 and 2019. Migration pressure will affect every country to a greater or lesser degree.

### LIFEBOAT "WE"

We can see something of what the future holds in roughly fifty years, in 2068, and probably much sooner if these trends continue—and it's certainly cause for worry. The slow, inexorable changes happening simultaneously in both biophysical and social systems, operating as tectonic stresses, are increasingly distending our institutions, economies, societies, and our global civilization; and when their pressures are released—the way pressures in Earth's crust are released in an earthquake—they may rupture our social systems. Even in wealthy countries, Ben and Kate's generation will likely experience conditions that are generally worse—and perhaps vastly worse—than those their parents experienced.

These changes are already producing disruptions in societies and people's lives that are making many across swaths of the world angry and afraid, even if people don't always recognize the ultimate sources of their discontent. They're eroding societies' social capital, which is the civility, trust, and unspoken commitments to fairness and mutual benefit that are crucial elements of well-functioning, peaceful communities. They're also making many leaders and governments seem ineffective and incompetent, which has diminished their moral authority—or "legitimacy"—in citizens' eyes and catalyzed anti-government sentiment, a dangerous move towards populism, and the rise of authoritarian leaders.

Rising anger and fear, eroded social capital, and diminished government legitimacy—this is an explosive slow-process mixture whose ingredients seem almost precisely selected to encourage social and political extremism. In many places, demagogues are already trying to ignite this mixture by pointing to scapegoats and encouraging their followers to defend "us" against "them." If they succeed, and we've seen it happen before, the slow rise of extremism could detonate in fast processes of social breakdown and mass violence.

It's a mixture that's also toxic to the basic values and principles that sit at the core of modern liberal society and are essential for democracy: freedoms of association, speech, and worship; the rule of law; and the basic human rights that protect us from the arbitrary use of power. In the absence of aggressive efforts in their defense, it's hard to see how anything resembling liberal society and a liberal international order can survive even through the next two or three decades, let alone till 2068.

What might arise in their place? We can imagine the emergence of a patchwork of insular and cruel authoritarian regimes that dominate their own territories, while competing for control—sometimes through war—of broad interstitial zones where a conjunction of tectonic stresses, weak government, ethnic strife, and economic and community disintegration has left behind varying degrees of anarchy and barbarism. Terrorist violence would likely seep from these brutalized regions into the lands of the remaining strong dictatorships. They'll respond by walling themselves off from the outside world even more and perhaps by deploying swarms of autonomous weapons beyond their borders to keep their enemies off-balance. What's left of human civilization would become progressively less connected and poorer, even as extremism, authoritarianism, and conflict cripple humanity's collective ability to address common—and still steadily worsening—problems like climate change.

The books and movies that fill our minds today with visions of future dystopia—*The Handmaid's Tale*, *Blade Runner*, *The Windup Girl*, *American War*, *The Hunger Games* and, again, *Mad Max*—may seem like nightmares, but the best of them are fully based on a reality that's already emerged or is emerging around us.[26]

All of which brings us to the techno-optimists' most fundamental error. While they're happy to talk about how things, up to now, have been getting steadily better for humanity as a whole—mostly referencing the past and often downplaying today's more ominous trends—and while some of them are pouring their considerable ingenuity and money into technological innovations, they're generally less keen to note that humanity's problems will only be manageable at a deeper social level, when humanity actually perceives itself, at last, as a whole—as an interdependent species confronting a common fate.

Humanity is already to a significant extent "one." We've created a tightly knit "socio-ecological" system that reaches into all corners of the planet.[27] Through trade, financial links, social media, infectious disease, global warming, and other phenomena, this system intimately affects the lives of people everywhere—country and city dwellers, farmers and factory workers, bureaucrats and bankers, poor people and rich. In such an integrated world, if we're to grasp the nature of today's emerging global dangers and adequately mobilize ourselves to do something about them, "we" needs to come to mean, to a lot of people most of the time, the entire human species.

Think about the lifeboat metaphor—commonly deployed but no less revealing for all that. Imagine all humanity has been crammed together into a small lifeboat that's being tossed by a huge storm on a pitch-black night. It's dark and chaotic. No can see clearly what's going on. Waves are crashing against the gunnels, and the boat has sprung leaks at one end. Some of us have done much more than others to create the leaks or have shirked the need for their repair; but through our actions together, advertent and inadvertent, just about everyone has contributed something to creating them.

Some optimists call out to the others not to worry: even though the water is already knee-deep at the leaky end, they declare that if the amount is averaged across all people in the boat, it's only enough to make everyone's feet wet. The false hope they offer makes it easier for the people farthest from the leaks to ignore a fundamental fact: if the leaks aren't patched, people farthest from the leaks will drown at best only a little later than those nearest the leaks. They'll drown even if they try to protect themselves by building barriers to stop the water surging towards them or by throwing some fellow passengers overboard. Meanwhile, the weight and pushing and shoving of most of the boat's passengers is making the leaks worse, as everyone expends their remaining energy arguing about who is responsible and who should therefore be first to repair them.

The leaks are an implacable equalizer: they create a common fate for everyone in the boat. And importantly, they can be patched only if people from all parts of the lifeboat work quickly and effectively together. As one.

"We" are everyone in the boat.

Today, creating such a species-wide feeling of "we-ness"—even if it's embryonic, or initially shared by only a fraction of people on our planetary lifeboat—is among our most urgent collective tasks.

A global identity is essential if we're going to find and implement solutions that will be enough to put us on a positive course into the future. It might seem impossible—our existing national, religious, linguistic, class, and ethnic attachments have such raw emotional power—and as fear ratchets up, these divisive attachments are gaining a stronger grip, at the expense of our sense of our common humanity. But in the next chapters I'll argue that a rapid shift towards a global identity is not only essential—it's entirely feasible, too.

IO

# A Contest of WITs

*The real problem of humanity is the following: we have paleolithic emotions; medieval institutions; and god-like technology.* E.O. Wilson

WE HUMAN BEINGS BRING VITAL STRENGTHS to the race to solve our critical problems.

For one thing, thanks to modern science, our species now has generally excellent knowledge about what our problems are and what's causing them. When it comes to climate change, for example, we do understand the physics and chemistry of Earth's atmosphere in detail (even if some folks don't want to accept this knowledge). We also have in place a global network of instruments that measures changes in the micro-concentrations of greenhouse gases such as carbon dioxide and nitrous oxide, a similar network of temperature gauges that shows how much the atmosphere is warming at Earth's surface, and constellations of satellites that tell us about the energy imbalance at the top of the atmosphere—between solar radiation coming in and heat and reflected light going out. We have similarly excellent knowledge about the many other problems we face, whether it's ocean acidification, threatened food production, mass migration, or the genetic makeup of the SARS-CoV-2 virus that causes COVID-19 disease.

This kind of knowledge puts us in an immeasurably better position than, say, Europeans were in the fourteenth century when confronted with another catastrophic threat, the Black Death, which went on to kill between 30 and 50 percent of the subcontinent's population. People at the time barely had a clue as to how their behavior and living conditions were helping spread the bubonic plague. But today, while scientists don't fully understand many things about Earth's climate, they do know the fundamentals of how we're causing it to change. They can also tell us what's likely to happen if we don't stop contributing to that change—if we don't modify our economies, energy systems, and the like. That's a fantastic advantage.

For another thing, great numbers of people everywhere are already pursuing countless projects with intelligence, enthusiasm, and often enormous courage to address our critical problems. Teachers in Bangladesh are educating girls to give them a means to economic independence, and with it greater freedom to control their lives, and when they have children; civic groups across southern Europe are coming to the aid of migrants arriving overland and by boat from Asia and Africa; environmentalists in Latin America are defending forest communities from illegal loggers; homeowners in Germany are installing solar panels and battery-storage systems to cut their carbon emissions; scientists are working in university laboratories to find potent new antibiotics to fight drug-resistant infectious disease; and children everywhere are organizing, with skill and tenacity, to demand climate and social justice and bring awareness of growing climate dangers. The list is long and credible.

And compared to even a few decades ago, our species now has a much greater ability to innovate in answer to our problems. That's because human social systems—whether community associations, municipal councils, societies, or planet-spanning institutions and corporations—are all instances of what scientists call "complex adaptive systems."[1] They learn and adapt by exploiting the power of combination—a power that's essential, as we've seen, to our recursive imaginations—as they join together bits and pieces of existing ideas, institutions, technologies, and practices in new ways to meet new challenges.

Our modern global connectivity gives us the potential to supercharge this "combinatorial innovation." It helps us link together in real time the world's staggering technical expertise, research capability, creative genius, and economic productive power, fashioning unprecedented collaborations among scientists, engineers, governments, firms, and citizen groups to develop new social arrangements and technologies in answer to our problems.

Think of molecular biologists who specialize in gene splicing coming together with physicists who are experts in fluid dynamics, biochemists who understand photosynthesis, and energy economists who work on transportation systems to develop ways to use algae to make massive amounts of carbon-neutral fuel for our cars, trucks, and airplanes. Or imagine architects working with materials scientists, urban planners, artificial intelligence specialists, lawyers, and artists to collaborate on breakthrough designs for super-energy-efficient, yet attractive and affordable urban villages. In some places, notably the world's great cities, just this kind of innovation is already flourishing in response to challenges like climate change.

In sum, humanity has within itself the knowledge, intelligence, enthusiasm, and economic resources it needs to solve its major problems. And, to be sure, in recent decades around the world, and especially in its poorest regions, we've been using these strengths to win the race to address crises like abject poverty, childhood infectious disease, and illiteracy.

But . . . but . . . even if one tallies up all the genuine successes to date and those that seem likely in the foreseeable future—as the techno-optimists do—the total is still not enough to address the full range of challenges our species is facing this century. Put differently, what humanity is doing now and is on course to continue doing won't get us to a future where we've significantly reduced the dangers arising from climate change, biodiversity collapse, worsening economic inequality, and the risk of nuclear war, to take just four challenges. Despite all our very real strengths, without radical change across all social systems, we're going to fall farther and farther behind in our race to solve many of our most threatening problems.

The campaign to stop nuclear testing can help us see why.

### HYSTERICAL DREAD

"Writing the letters made me feel better," Stephanie May wrote in her memoirs. She was referring to her October 1956 campaign to convince influential women to support the Lobby for Humanity to stop nuclear testing. "I was young and naïve enough to believe the idea might actually catch on." But the replies that arrived in her farmhouse's postbox proved disheartening.

> Chartwell,
> Westerham,
> Kent.
> 14th November 1956
>
> Dear Madam,
>
> Lady Churchill has asked me to thank you for your letter of November the 8th, and to say that while she can well understand how you feel in the matter, she does not

wish to lend her name or join the organization which you think of forming.

Yours very truly,

G. Hamblin
Private Secretary

Clementine Churchill, Winston Churchill's wife and a beacon of strength in England through World War II, clearly wanted nothing to do with Stephanie's proposal. Eleanor Roosevelt answered that "your idea seems to me a very ambitious one, but I doubt, unless this is done through government, that it can be done at all."[2] With one exception, similar responses came from all "who bothered to write back," Stephanie observed.

That exception, though, made a huge difference. Arriving in mid-November, it was a handwritten note from Elizabeth Ives, an author and the sister of Adlai Stevenson, the Democratic candidate in the just-concluded US presidential election. Stevenson had made banning atmospheric nuclear tests a central plank in his unsuccessful campaign against the incumbent Republican president, Dwight Eisenhower, despite warnings that the political cost would be severe.

Still, just ten days after an election result that Elizabeth Ives must have found devastating, she answered Stephanie's October letter: "I think your idea of a 'Lobby for Humanity' has true merit and inspiration," she said. But "you will have to work hard and organize it well," she advised, offering her help and her name in support. When Stephanie replied with news of the disappointing response to her campaign overall, Elizabeth urged her not to be discouraged. "No good idea is ever lost if it is persisted in, even by one follower. . . . Events are moving fast, and it may be that with your continued persistent efforts towards publicity, you may get more success than you realize."[3]

In late December, another reply appeared from the journalist, academic, and historian Gerald W. Johnson. It was gracious and supportive. He said he thought a US moratorium on nuclear testing was imminent, but he also offered some words of caution: "As for the organization of women you propose, it will be extremely difficult to create because of the hysterical dread of anything that could be made to seem friendly to Communism, and this could."[4]

Johnson had put his finger on a critical obstacle facing any American who wanted to stop nuclear testing: in the mid-1950s, a large majority of people thought their country was locked in an epic, Manichean struggle with Soviet communism, a struggle that would ultimately determine both national survival and humanity's future. Many of them believed any hint of weakness was abhorrent; and since halting testing could signal weakness, they often branded anyone advocating a halt as either stupid or a communist sympathizer.

Even so, Johnson, by highlighting only a single set of beliefs (anti-communism) held by many people in his society at the time, radically understated the breadth and depth of the obstacles that Stephanie and her fellow activists faced, and why positive social, political, and economic change was so hard to achieve then—as it is now for all the groups mobilizing to fight problems like the climate crisis. A new theory that's grounded in complexity science helps us see the larger picture—and understand why.

## WIT CLICK

Every year I read hundreds of academic articles. Most are useful and interesting, but only a few tell me something truly new. And only the very rare article—maybe one in a thousand—leads me to see things in a profoundly new way. In those infrequent moments, something goes "click" in my mind, as if the pieces of a puzzle have suddenly fallen into place to reveal a larger pattern.

In 2009, I had this reaction to an article published in the US journal *Proceedings of the National Academy of Sciences*. It was unexcitingly titled "Overcoming Systemic Roadblocks to Sustainability: The Evolutionary Redesign of Worldviews, Institutions, and Technologies." The ecologist Rachael Beddoe, economist Robert Costanza, and eleven co-authors proposed that human societies are organized around cohesive sets of worldviews, institutions, and technologies.[5]

The article's authors stress that these particular three components—worldviews, institutions, and technologies, which they labeled rather sweetly as WITs—are tightly interdependent: they influence each other, depend on each other, and usually hang together in a cohesive way.

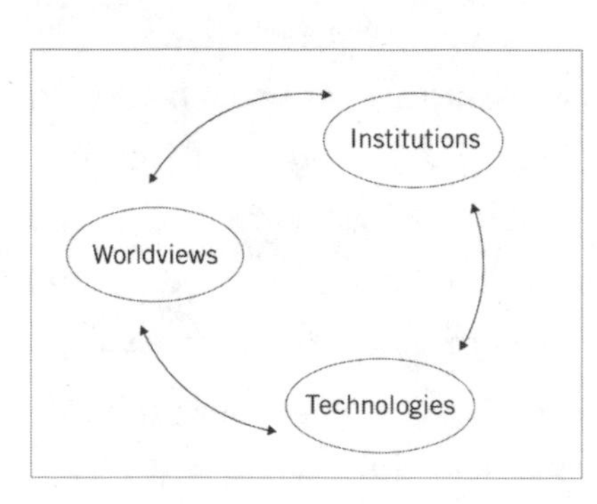


So what does that mean? Here's an example: a prominent part of the Western worldview is a commitment to personal liberty, independence, and free enterprise. This commitment supports—and is supported by—the institutions (banking and monetary systems, property and contract laws, and so on) that create our (supposedly) free economic markets, which themselves in turn both channel and enable our freedom. The commitment to freedom also supports—and is supported by—the technology of private cars (as one example), because cars let us zoom around feeling free, at least until we're stuck in traffic with everyone else who's trying to zoom around.

The links among these three WIT components have real, practical consequences: they mean, for instance, that we're going to find it hard to get people out of their cars or, more profoundly, change

the way our markets work without also addressing people's beliefs and feelings about their personal freedom.

*Worldviews*, as I said earlier, are mental networks of concepts, beliefs, and values—often emotionally charged—that allow us to interpret our circumstances and plan our actions in response. They usually give our lives meaning and therefore some sense of security, so they're fundamental to our sense of self through time, and they're at the heart of our capacity to adapt to changes in our circumstances.[6] *Institutions* are, broadly, a community's rules, usually communicated and learned informally from our families, peers, and social groups or formally in places like schools.[7] These rules range from the mundane (those that tell us on which side of the road we should drive) to the vastly more complex (like the laws governing markets and parliamentary democracy). They also include a culture's subtle and unwritten social norms about what behavior is appropriate or ethical at specific times and places. Finally, *technologies* are problem-solving tools that we create using energy and information to exploit properties of our physical environment.[8]

Worldviews, institutions, and technologies play different roles in our lives. Worldviews directly affect what and how we think, while institutions and technologies directly affect what we do and how we do it. And it's a perpetual circle. We create our worldviews in a multitude of conversations with other people; sometimes these conversations are direct, but more often today they happen indirectly through technologies such as books, movies, and smartphones. In the process, we create a scaffolding of shared ideas through which we can, together, invent and deploy more complex institutions and more technologies. These institutions and technologies in turn then powerfully shape our worldviews, especially how we categorize people around us and what possibilities we can imagine and desire. This process creates a "WIT."[9]

What really struck me most in the article was its novel argument that WITs are the primary "unit of selection" in the evolution of our societies. In other words, by carrying a society's information and structure *through time*, a dominant WIT plays a role in society analogous to a gene in a biological system. Whether the WIT survives or is replaced by a less dominant WIT depends on the "fitness" it confers to its social system—that is, on the extent to which it helps that system reproduce itself in its larger environment.

This is a truly penetrating insight and a vital addition to our picture of the causes of humanity's global crisis. It suggests that the crisis results from a worsening mismatch—or a lack of fit—between the specific WITs that currently dominate our societies, on one hand, and the fundamental properties of the global socio-ecological system in which our societies are now embedded, on the other, particularly that system's increasingly degraded natural or "material" components.

Put in everyday terms, our societies' prevailing beliefs and values now are too self-centered, our political systems too hidebound and shortsighted, our economies too rapacious, and our technologies too dirty for a small, crowded planet with dwindling resources and fraying natural systems. It looks increasingly likely that today's dominant worldviews, institutions, and technologies will soon lose the unforgiving evolutionary contest to other WITs, still to be determined, that are better-adapted to our ever more extreme situation.

As evolutionary theorists put it, they'll "be selected out," which is a euphemism for "go extinct."

### SOCIAL POWER

We can take the arguments of the article's authors further. Perhaps because they are mainly ecologists and economists, they don't really discuss how every WIT is embedded in, structured by, and sustained through systems of social power. Nor do they discuss how every WIT is also a conduit through which some people and groups exercise their power over other people and groups. But these two mechanisms are an essential part of our story, because our societies' worldviews, institutions, and technologies both circumscribe and create people's opportunities.

Social power is among the most elusive phenomena studied by the social sciences.[10] Nonetheless, it pervades all human systems, often unseen but never idle. Social power is far more than the ability of one person or group to get another to do something they otherwise wouldn't, which is the way we usually think about it. It also operates through the differing options and opportunities available to people and groups. Basically, as we all recognize, if you've got less power, you've got fewer opportunities—and vice versa.

More subtly and subversively, social power also inheres in the very language we use to communicate with each other, to describe and to manipulate the world around us, and to build our own worldviews. It suffuses the words we use to categorize things and people, the sentences we use to define relations between them, and the stories we tell to include or exclude people from our communities and their decision-making. And it suffuses, too, the concepts and beliefs we use to define what counts as a problem and what doesn't.

Stephanie clearly grasped that she was struggling to change a formidable cluster of interlinked worldviews, institutions, and technologies—one consisting of far more than simply entrenched anti-communism. On the opening page of her first scrapbook is a pen-and-ink cartoon by Herb Block, winner of three Pulitzer Prizes in his long career (and known as Herblock). It was first published in May 1957 in the liberal magazine *The New Republic*, alongside an article by two scientists on the hazards of nuclear tests.[11] The cartoon is a mordant and bitterly insightful statement that captures all three components of what one could call the era's "nuclear state-system WIT."

Block represents nuclear-weapons technology as a thuggish oaf drawn to resemble a crude iron bomb or artillery shell, with an unlit fuse sticking out from the top of his head. But notice a subtle detail: the oaf's muscular arms and thick hands contrast with his crisp, buttoned shirt cuffs, suggesting he's wearing a business suit or a lab coat. Block is hinting at the links between the brutal violence implicit in nuclear weapons and what President Eisenhower would later call—in his 1961 farewell address—the "military-industrial complex" and the "scientific technological elite."

This brutish techno-institutional system is, in Block's cartoon, leashed to a silly little man who looks like a circus clown. He's labeled "Atomic Energy Commission," which was then the US federal agency responsible for nuclear weapons development and regulation. The little man stands for the institutions of the state and

NUCLEAR FALLOUT DANGER
ATOMIC ENERGY COMMISSION
HERBLOCK
"LOOK, LADY – YOU DON'T SEE ME WORRYING"

government, but his leash is thin and crudely knotted: the laws and regulations intended to control this cruel technology are pathetically weak. While the big nuclear oaf juggles a baby with his eyes closed—implying that the risk to kids from radioactive fallout is enormous but random—the little man declares, to the child's terrified mother below, his blithe and ignorant indifference. The mother could be Stephanie May, to whom Block gave an original rendering of the cartoon.

It's all there in one amazingly deft drawing: the interlocking WIT elements of nuclear-weapons technology; of bureaucratic, scientific, and economic institutions; and, most subtly but perhaps most tellingly, of a powerful group's beliefs and values regarding the relative authority and importance of the state, nuclear technology, and mere moms and kids.

Perhaps surprisingly, the fear of communism doesn't figure anywhere in the image, even though Block himself coined the term "McCarthyism" in scathing cartoons drawn earlier in the decade, depicting Senator Joseph McCarthy's efforts to root suspected "Reds" out of the US government and Hollywood. It seems he intuited that, in the complex system of worldviews, institutions, and technologies surrounding nuclear weapons, much more was at work than anti-communism.

# 11

# Why Is Positive Change So Hard?

*Governments do not like to face radical remedies;*
*it is easier to let politics predominate.* Barbara Tuchman

SYSTEMS OF SOCIAL POWER SURROUND and sustain our societies' dominant sets of worldviews, institutions, and technologies (WITs). That's a good starting point for gaining a better understanding of why major, positive change is so hard to achieve—and why we're falling ever-further behind in our race to solve many of our most dangerous problems. But we need to understand other key factors, too.

We've already considered how change sometimes doesn't happen because people aren't aware it's needed. If critical problems aren't readily visible—as is true today for worsening antibiotic resistance—then most people won't realize anything needs to be done. Stephanie May faced this challenge, because nuclear testing's radioactivity was invisible and its health effects random and slow to appear. Activists today like Greta Thunberg face the same problem to some extent: climate change is becoming more apparent to people, but many of its impacts are still hidden below the ocean's surface, in the atmosphere, or in remote ecosystems—and so not widely recognized or comprehended yet.

When a previously unseen problem becomes somewhat more visible—as when climate change starts to cause surprisingly powerful storms or wildfires—we understandably become frightened. One might think that this emotional response would motivate change, and sometimes it does, but it can instead trigger psychological defenses—like various forms of denial—that keep people from acknowledging the problem's urgency. This reaction is common when people think they can't do anything about the problem in question. And the sheer size and scope of many of the challenges that humanity faces today accentuate our sense of impotence by obscuring what should be done and who should do it.[1]

Take climate change. It arises not just from our everyday activities—like driving gasoline-powered cars, eating meat, or, in poorer countries, using cooking fires—but also from the activities of heavy industries consuming carbon-based fuels; of governments, when they subsidize fossil-fuel industries; and of international trade organizations, when they restrict policies like national carbon tariffs on imported goods. So humanity must act simultaneously across many, often radically different social, economic, and political domains, and across many "scales"—from the household, community, and city all the way up to the national and even global scales. That makes coordination and cooperation around solutions far harder; and since nearly everyone and everything is implicated, it's easy to think one can't do much alone and then point a finger at someone else to start first.

Yet even when it's more or less clear what should be done about a problem and by whom, the people, groups, or corporations with the means to find and implement solutions often don't have incentive to do so—because they don't think it will pay them to do so.

Economists have a name for this situation: "market failure."[2] The concept assumes that when a market is working well or "efficiently," the prices of goods and services traded in it will reflect the full social costs and benefits of producing and consuming those goods and services. Prices often don't reflect these costs and benefits, economists say, when property rights are either unclear or hard to establish. If something is to be bought and sold in a market, its ownership must be clear and transferable.

When markets fail, good solutions to problems are undersupplied in both quantity and quality. Because the prices of fossil fuels that we use to heat our buildings, power our cars and trucks, and generate much of our electricity don't currently reflect the harm those fuels' use inflicts on the climate and the well-being of current and future generations, they're still relatively cheap compared to less dangerous alternatives like electricity from the sun or wind. This of course weakens consumer demand for those less dangerous alternatives, which then lessens the incentive for conventional energy companies to develop and market them. Likewise, because the prices of robots used in manufacturing don't reflect the broader social costs of the job loss that the devices cause, manufacturers have less cause to invest in other ways of boosting their output that would use more labor. Or sometimes the people who'd benefit most from viable solutions to a problem don't have the money, or can't coordinate themselves, to pay the corporations who could supply the solutions. For instance, poor farmers in tropical countries can't pay enough for new grains that can tolerate drought and heat shock, so multinational agricultural companies don't invest enough in developing such grains, despite the looming climate crisis for agriculture.

It's a grim reality today that many of humanity's emerging global problems—like climate change and biodiversity loss—produce the conditions for market failure in stunning abundance. These problems spill across geographical and political boundaries, so they don't fall within any country's markets or regulations. And they tend to affect public goods or complex resources like Earth's atmosphere, for which it's extraordinarily hard (perhaps happily) to establish property rights—something that would push profit-driven corporations to take an interest. So good solutions to such problems are woefully undersupplied.

The job of fixing market failures usually falls to national governments.[3] They can try to invent and deliver solutions themselves—developing new, more sustainable grains and fuel in government-funded labs, for example. They can also increase subsidies to companies to find and deliver solutions—like renewable energy—while cutting subsidies to, say, fossil-fuel companies that are making climate change worse. Or, they can make markets function better by coordinating buyers and sellers; setting or clarifying property rights, as they sometimes have with cap-and-trade systems for carbon dioxide emissions; or changing incentives through devices like taxes on carbon—or robots. In other words, governments can, in theory, work to solve a problem—market failure—that's preventing people, groups, and companies from solving other problems.

## THE SOCIAL POWER OF VESTED INTERESTS

But here's where we run headlong into perhaps *the* key factor that hinders the supply of solutions to our critical problems and, ultimately, makes positive change so hard to achieve: the blocking power of strong vested interests who want to maintain the status quo. Governments often avoid putting in place the policies, regulations, laws, and institutions that would bring needed change, because vested interests undermine or reject them outright. This is one of the conduits through which social power has its greatest effect in our societies.

Social scientists define vested interests as groups focused on gaining benefits for their own specific segments of society. Of course, they come in many forms, and one person's ignoble vested interest is another person's noble and vital protector. A mining company might save its employees' jobs by lobbying against environmental regulations; a union might stop a company from gutting its employees' pension plan by striking; and a religious community might have extra resources to bring comfort to its members by gaining a tax break. How one views these pressure groups depends on where one stands, because, again, what counts as a problem—and therefore what counts as nefarious action to block a solution to that problem—depends on one's values, goals, needs, and circumstances. There's no entirely objective test for venality, or for altruism, for that matter.

Yet the actions of vested interests create a broad bias in societies towards stasis, tending to strengthen already entrenched structures of social power and reinforce already dominant sets of worldviews, institutions, and technologies (WITs).[4] These groups generally use every tool available—legal and, sometimes, illegal—to defend the benefits they derive from the existing ways of doing things. They exploit people's fears and, through massive political contributions to sway elections and otherwise, torque governments into maintaining streams of subsidies and regulatory advantages. Too often, when it benefits them, they promote lies: fossil-fuel companies, for example, have spent millions upon millions of dollars propagating phony science on global warming, just as tobacco companies earlier spent fortunes trying to discredit studies showing a link between smoking and cancer, all with the aim of dismantling existing or preventing new regulations.

In the Herb Block cartoon shown in the last chapter, the colossal vested interests of the military-industrial-techno-science complex tower over a Stephanie-like figure representing mothers everywhere. This complex, which President Eisenhower declared held "the potential for the disastrous rise of misplaced power,"[5] amplified people's fear of the Soviet threat to secure its economic and political position in US society. I suspect that's why Block didn't depict anti-communism—a political view broadly held in US society at the time—anywhere in his cartoon: he recognized that these beliefs were, in many respects, largely a consequence of and cover for the real mechanisms of power and influence driving nuclear testing.

The actions of vested interests degrade the quality of governments' solutions to problems and make anything but incremental change almost impossible. When a government faces a new problem, or an old problem that's become so bad it can't be ignored, it hardly ever scraps its current way of doing things to create a new and better approach. Instead, it usually adapts an existing procedure, institution, or policy to address the problem, or it adds a new procedure, institution, or policy on top of existing ones.

Exhibit A is the response of the world's nations to the 2008–09 financial crisis and its aftermath.

In the wake of the crisis, because of lobbying by powerful financial interests, major governments and their finance ministries left in place almost all the financial system that had contributed to the disaster, including large banks, insurance companies, ratings agencies, and national and international regulatory regimes (or at least those parts of the system that hadn't gone bankrupt, merged, or otherwise vanished because of the crisis itself).

It's true that after the crisis some actions were taken. The United States, for instance, passed the Dodd-Frank legislation that imposed new rules on banks' capital requirements and greater government scrutiny of bank lending, while reforming some oversight bodies and adding others. The governments of the world's biggest economies also set up the Financial Services Board, which among other actions produced a list of "Systemically Important" (or "too big to fail") financial institutions;[6] and the international Basel III Accord established new voluntary rules for capital requirements to protect banks against runs.

But in the United States, the new rules evoked howls of protests from the financial sector, and in 2018 they were partially rolled back by the Trump administration. And none of these actions, whether national or international, ever changed the basic structure of the global financial system and its core actors. Most importantly, by designating certain banks and institutions as too big to fail, rather than just breaking them up, governments in some respects entrenched even greater risks in the world financial system. Now, because the designated companies know that the governments will bail them out, come what may, they have more incentive to gamble (an outcome known as "moral hazard"). Rather than reducing the underlying problems that caused the original crisis, in other words, government responses largely papered over those problems and may have even made them worse.

So old procedures, institutions, and policies, even ones that don't solve problems well, tend to stick around, only to be made steadily more complex with ad hoc tweaks and adjustments, while new procedures, institutions, and policies accrete layer on layer, atop the older ones. Eventually, the whole mess becomes sclerotic, opaque, and uncreative—and even less effective at solving the problems it's supposed to solve.

Social scientists have studied how the influence of vested interests has contributed to ineptness in bureaucracies, the failure of large corporations, the decline of nations, and even the collapse of history's greatest empires.[7] Today, vested interests are one of the major reasons why our current ways of doing things—our dominant

WITs, in other words—are so firmly locked in place, and so hard to budge, even in the face of grave threats.

### OUR COMMITMENT TO GROWTH

By far the most powerful set of worldviews, institutions, and technologies in our world today revolves around economic growth—a decidedly Janus-faced phenomenon that increases the well-being of many people the world over, but only at enormous cost to nature and risk to humanity's long-term health and even survival.

Humanity's commitment to "growth"—generally defined as a steady increase in our economies' output of goods and services, which supposedly brings greater abundance and diversity of desirable stuff to consumers—is reflected in the attitudes and aspirations of people almost everywhere on the planet, in the goals of just about all companies and public economic institutions like central banks, and in the policies of virtually all governments, from municipal councils to national bureaucracies, whether they're ideologically conservative, socialist, neoliberal, or protectionist.[8] This commitment is reinforced by the power of capitalist elites, and it's enabled by a constant stream of new technologies that produce wealth ever more efficiently from inputs of capital, labor, and natural resources. Almost no one, left or right, really wants to give it all up.

Nestled inside the worldview component of this "growth WIT" is the dominant intellectual rationalization of today's world order: conventional economics. This elaborate apparatus of theory, empirical evidence, mathematical gymnastics, value judgments, and self-congratulation legitimizes globalized capitalism and the social power of its elites.

At its core are the implicit assumptions that our economies are separate from nature and operate much like machines, so their behavior is mostly linear and potentially predictable.[9] If experts can discern underlying universal economic laws at work, this perspective's logic continues, then these economies can be managed by a planet-wide class of technocrats, including central bankers and government officials—a priesthood trained in the arcane science of economics.

Over the last two centuries, and particularly since the shocks experienced in World War II, the growth WIT has been at the heart of the processes I described earlier: our societies have used economic growth to deal with their visible and immediate problems, often including poverty. But that growth has then generated a raft of less visible and immediate problems, such as greater air and water pollution. Sometimes we've then used the wealth derived from growth to try to deal with those problems too, when they've become visible enough to be a concern; for instance, in the case of pollution, we've invented and deployed cleanup devices like smokestack scrubbers and sewage treatment plants.

But continued economic growth, at least at a scale resembling what we've known to date, is now fundamentally incompatible with the radically altered material world around us—a world of climate change, collapsing biodiversity, and dwindling natural resources. So the growth WIT is going to have to change drastically—or it, and possibly we, will vanish. Either way, the results for our societies are going to be wrenching.

Let's walk through a simple example of this incompatibility. If we take the average economic growth rate of the world as a whole since the 1960s (about 3 percent) to indicate what we can expect through this century, then by 2100 the global economy will be more than ten times larger than today's. Gross world product (GWP) was about $80 trillion in 2018. An average real growth rate for the global economy of 3 percent annually generates a GWP of slightly over $900 trillion in today's dollars by 2100.

Even if we're able to cut our carbon emissions per dollar of economic output by 80 percent, perhaps through heroic applications of new energy technologies, overall growth at the rate indicated will still double the world's already extremely high emissions. Doubled emissions would give a huge extra push to climate change, causing such extreme heat waves, droughts, and storms that farmers almost certainly won't be able to feed the world's ten-billion-plus people. Climate change creates hungry people, hungry people create political instability, and political instability is growth's worst enemy. In short, the environmental, and in turn political, damage that growth will cause will eventually halt growth.

Hold it, you might say. Surely, we can have both economic growth and a clean, stable, and healthy natural environment too? Why can't we cut carbon emissions to zero even with a growing economy? That's the whole premise of "sustainable" growth, isn't it, where the right kind of growth needn't sacrifice the well-being of future generations?

The idea of sustainable growth, often referred to as "sustainable development," has been around since the 1980s, when it was introduced in the Brundtland Commission's famous report *Our Common Future*. Widely published in 1987 and named after Gro Harlem Brundtland, then Norwegian prime minister and chair of the World Commission on Environment and Development, the report said sustainable development "seeks to meet the needs and aspirations of the present without compromising the ability to meet those of the future."[10] The idea of sustainability is now a mantra of today's economic policy elites, corporate leaders, finance ministers, and the like. To the extent that these folks acknowledge Earth's environmental constraints at all, when they speak of sustainability they're claiming, essentially, that we can have our cake and eat it too. Yet look closely, and one finds that the idea of sustainable growth rests on a cluster of profoundly flawed premises.

**Getting the Prices Right**

The first premise is that protecting the environment is simply a matter of dealing with market failures. If people pay the right prices for environmental goods and services—by "internalizing" costs that are otherwise "external" to market prices, as economists put it—then they'll have greater incentive to conserve them. Residents of my hometown must pay a fee for every bag of garbage municipal workers collect, to cover the cost of managing the landfill; so, most homeowners are careful about how much stuff they throw out. But people around the world pay little or nothing to use the atmosphere as a garbage dump for their carbon dioxide. If, instead, we were asked to pay the costs that our children and grandchildren will bear from the climate change that our carbon output today will cause, we'd emit a lot less carbon.

There's nothing wrong with the economic principles behind the idea of internalizing costs. Still, when we come to implementing these principles, many devils haunt the details. For instance, we can't be precisely sure about the future costs of climate change, and we must make some critical assumptions about how much we're prepared to pay in the present to avoid those future costs (the question, as economist call it, of the "discount rate"). Also, things like pollinators, fish of the high seas, and Earth's atmosphere are highly dynamic "complex resources" (chapter 3). These and other factors mean that applying property rights to our natural environment or otherwise pricing its goods and services "correctly" requires elaborate regulations and institutions. In their absence, these very special goods and services tend to be underpriced and overused.

Sadly, experience over recent decades around the world shows that powerful vested interests always impede, torque, or hijack government efforts to "get the prices right"—that is, to ensure that people pay a larger fraction of the full social costs of using the natural environment. Even though economists widely agree on the basic principles behind carbon taxes and cap-and-trade policies, for example—and argue forcefully in favor of such policies—and even after decades of academic discussion, countless learned policy briefs, and some half-hearted government moves in Europe, Canada, and China, we're still not paying anything close to an appropriate price for throwing our carbon junk into our air.

**The Environmental Kuznets Curve**

A related argument often used to defend the idea of sustainable growth is based on what's called the "environmental Kuznets curve," after Simon Kuznets, a renowned twentieth-century American economist. It says that as poor societies become wealthier, the amount of damage they cause to the environment at first increases but then at some point starts to decrease.[11] So a graph of how environmental damage changes (on the vertical axis) as wealth goes up (left to right on the horizontal axis) shows a curve first rising and then falling, like an inverted U.

The curve's implication is that we can fix our environmental problems by getting rich as quickly as possible. And when we look at the historical data for wealthy democracies in Europe, North America, and Asia, this inverted-U relationship has in fact held for some pollutants like sulfur dioxide, lead, and urban smog and sewage. As these countries' citizens became richer, they turned their attention from day-to-day survival to, among other things, the state of their environment. And they said to their leaders: "Fix this mess. It could be making us sick. We don't like the look of it. It smells. Clean it up!" Then these countries used some of their great wealth to buy "end-of-pipe" solutions such as scrubbers and treatment plants.

This is a classic example of the phenomenon where people deal with environmental problems that are palpable in their air, streams, rivers, and lakes, in the process making worse less visible and immediate ones. Almost always, the end-of-pipe pollution controls we adopt involve higher energy use, and because much of our energy still comes from fossil fuels, higher energy use usually boosts carbon dioxide output. The catalytic converters we started sticking at the end of our tailpipes in the 1970s lower our cars' overall gas mileage by about 10 percent. The result was less visible smog but *more* invisible $CO_2$.

Basically, we took advantage of a market failure—inadequate pricing of $CO_2$ pollution—to transfer pollution costs beyond our horizon in time and space, so they'd be borne by people elsewhere on the planet—especially those in poorer countries with less capacity to adapt to climate change—or by people elsewhere in time, like our kids.

In fact, contrary to what enthusiasts for the environmental Kuznets curve say should happen, the $CO_2$ output of most countries is still climbing steeply as they get wealthier. For instance, between 2010 and 2018, Turkey's $CO_2$ output per person rose 21 percent and India's a stunning 43 percent (albeit, in India's case, from a 2010 level less than one-tenth of Canada's).[12] And while it's true that emissions per capita are starting to level out and even decline in the world's richest countries—in Germany, for instance, and the United States—there's an essential caveat: the peak is happening at extremely high levels of emissions and wealth, and the decline afterwards, to the extent that it's visible at all, is halting and slow. So far, at least, the last half of the inverted U—the downward part—is largely missing. And globally, we don't yet see anything resembling the full environmental Kuznets curve; the 2018 jump in global emissions from fossil fuels was 2.7 percent, more than twice the rate of humanity's population growth, and the largest in seven years. (Global emissions declined markedly in 2020, as the COVID-19 pandemic caused a sharp economic contraction, but the decline simply reflected the tight short-term connection between emissions and economic output.)

**"Green Growth"**

But maybe we're just not trying hard enough. Maybe if governments and companies energetically collaborate to focus our immense wealth and technological prowess on finding ways to make our economies greener and more efficient—deploying everything from better lightbulbs to mag-lev trains—we can, country by country, push carbon emissions and other kinds of environmental damage and resource depletion towards zero, even while economic growth continues. Progressive, market-oriented environmentalists use this third argument to back the idea of sustainable growth, which they usually call "green growth."

Good evidence supports this argument in favor of sustainability, so much so that it's the one I find most persuasive—up to a point. Since the oil shocks of the 1970s, rich countries have achieved huge gains in efficiency, defined in terms of the amount of energy or material they use—or carbon they emit—per dollar of GDP they produce. Rates of improvement of 2 percent annually have added up fast. In the United States between 1980 and 2014, for example, the amount of energy used per dollar of GDP was cut in half.[13]

But once more, there are caveats. For one thing, only about 60 percent of this improvement in the United States was due to real efficiency improvements in US production, through better factory technology, for instance. The rest came from the US economy's transition towards less energy-intensive industries, which means some of the economy's energy use was displaced overseas. Basically, the US economy became somewhat cleaner because some economies overseas became dirtier. For another thing, overall economic growth swamped a lot of the US efficiency improvement, so by many measures its total negative environmental impact rose steadily. Between 1980 and 2014, energy efficiency gains in the United States were overwhelmed by real economic growth of 2.75 percent a year, so the economy's total energy consumption still went up by a quarter (from 78 to 98 quadrillion BTUs per year).

On balance, the historical track record shows that we can't come close to solving growth's environmental impacts through better technologies and efficiencies alone.[14]

**"We Can Fix the Problems Later"**

And then there's the fourth and final argument that's sometimes used to back the idea that growth can continue indefinitely—and it's by far the worst of the bunch. It says we can delay fixing our environmental problems until they get atrociously bad and then simply use all the wealth we've generated in the meantime—by trashing our house, so to speak—to fix the environment. This argument is a kind of crude techno-optimism: it assumes that human beings are ingenious enough to find artificial substitutes as needed for every good and service that nature provides—chemical fertilizers for depleted soil nutrients, irrigation systems for absent rainfall, and mini-robotic insects for extinct pollinators such as bees.[15] It also ignores the very real possibility of hysteresis in these systems—that is, for irretrievable flips to new states with possibly catastrophic results for human beings.

When the Brundtland Commission introduced the idea of sustainable growth in the 1980s, it was worth taking seriously because we didn't know yet—not widely, anyway—that its underlying premises were so flawed. But in the decades since, we've learned through hard experience that sustainable growth is at best a fantasy and at worst a bald-faced lie—a pernicious source of false hope.

## THREE EQUIVALENCIES

When large numbers of smart, well-intentioned people delude themselves like this, it's usually a sign that something else is going on. So what powerful underlying motivations and beliefs are encouraging us to twist reality into a pretzel?

I think two general factors are at work. First, such a commitment to and belief in economic growth complements other worldview commitments that are widely shared by our world's economic elites. Growth fits with an optimistic, individualistic creed that encourages change, looks to the future, exults agency, and holds that getting rich is good. Second, a lot of hard, real-world evidence shows that people and societies gain many benefits from continual growth. Time and experience have distilled this evidence into three assumptions about growth's benefits that are now deeply rooted in our dominant economic worldviews. I call these assumptions the "three equivalencies."[16] They are:

**GROWTH EQUALS HAPPINESS.** Most people unconsciously (and sometimes consciously) associate more money with greater happiness, so it's widely assumed that since growth raises average incomes, it makes people happier. Yet the research disputes a straight correlation between income and happiness. While the correlation is strong at lower incomes, it may weaken as incomes rise and as people's basic needs are satisfied. Some studies suggest that beyond a certain point, thought to be around an income of $15,000 per person per year, each increment of extra income produces a smaller return than the last, and relative income—that is, whether one makes more or less than the next gal or guy—seems to count more towards happiness than absolute income.

But let's be clear: even if this research is correct, enormous numbers of people in the world still fall below this threshold. For the 3.5 billion people living on $5.50 a day or less (as of 2015, in 2011 US dollars), higher incomes from economic growth are essential to meet the basic requirements of human dignity; while for the 10 percent of the human population still living on $1.90 dollars a day or less, higher incomes can make the difference between life and death.[17] For them, the link between growth and happiness is indisputable.

**GROWTH EQUALS FREEDOM.** History shows that societies with economies that don't grow tend to become sclerotic. The inhabitants of pre-modern agrarian empires, for example, had few opportunities to change their status and economic and social roles. Instead, institutions or hereditary elites assigned people their roles, which were then largely fixed for life. Opportunities for individual expression were also narrow. But more recently in Western societies, economic growth has combined with modern notions of social, intellectual, and scientific progress—ideas rooted in the Renaissance and Enlightenment—to open vast space for personal liberty. Some conservative scholars even argue that people can't be truly free in economies that don't grow.[18]

**GROWTH EQUALS PEACE.** This is the big one. Humanity learned a brutal lesson in the 1930s: economic crisis and the unemployment and social dislocation it causes can nourish political extremism and contribute to horrible outcomes like World War II and the Holocaust. The economist John Maynard Keynes understood this relationship and in response invented theoretical and policy tools designed to maintain perpetual growth by sustaining economic demand. In 2008–09 and again in 2020, the world's central bankers and policy makers used every Keynesian tool they could grab to prevent global economic collapse, because they feared the social crises that come with soaring unemployment and economic loss. (How ironic that Keynes, an icon to many on the liberal left, may have provided the intellectual justification for policies that are helping to wreck Earth's environment.)

A profound contradiction—one with near-syllogistic simplicity—reinforces this last equivalency that growth equals peace. In capitalist market economies, the fight for survival and for profit among companies drives them to search incessantly for methods to lower costs of production. Many of these methods, including better technologies, replace workers, so raising unemployment. Yet the overall economy needs employed people to *buy the output* of the ever more productive companies. This is a recipe for both constant economic crisis and, as unemployment rises, political upheaval.[19]

Economic growth then rides to the rescue again. It creates new industries that absorb workers displaced by technological change, and these employed workers then provide the economic demand that keeps the economy humming. But note that growth doesn't really resolve the underlying contradiction: it just temporarily eases its ugly consequences until the cycle starts over.

MANY NOW REGARD THESE THREE assumptions about growth's benefits as virtually unassailable—as rock-solid truths—and not unreasonably so, given the evidence.[20] They also underpin a certain kind of hope: for those of us at the bottom or, for that matter, the middle of the economic hierarchy, growth is usually associated with social mobility and hopes for a better life in the future. Halting growth would freeze current inequalities in place. So, humanity faces a terrible dilemma: we can't live with growth as we've known it, yet at the same time we really can't imagine living without it—or even imagine wanting to live without it.

The mismatch between humanity's growth WIT and the reality of our degradation of the planet—as humanity's population soars towards ten billion and beyond—is our biggest problem of all. In some ways it encompasses and subordinates all other global problems, including climate change. Our political, business, and intellectual elites almost completely ignore it.

But, as we know, ignoring problems rarely makes make them go away. Today's conventional growth can end voluntarily, if we deliberately move the global economy onto a new path through economic and social innovation (of which more later); or it can end involuntarily—probably with social catastrophe in its wake.

But either way, it will end.

Indeed, it may already be starting to end. Northwestern University economist Robert Gordon argues persuasively that the impact of the wave of disruptive technologies since the late nineteenth century has now mostly passed through the richest economies of the world, and he doesn't see any technologies on the horizon that could produce the same broad growth effects.[21] Many of the new technologies in the pipeline today—autonomous vehicles, artificial intelligence, advanced robots and the like—appear likely, on balance, to substitute capital for labor, except for labor requiring very high cognitive skills. This means that they'll generally worsen problems of unemployment, economic inequality, and therefore economic stagnation. So, over time, "real" growth in rich economies—that is, growth after inflation is subtracted—could trend down towards the long-term historical average, prior to about 1850, of less than 1 percent annually.

And what growth various economies experience will also have to overcome several "headwinds," including worsening economic inequality, high public and private debt, and rising costs for energy and from environmental damage. Climate change, especially, is already substantially retarding growth—to the tune of trillions of dollars of lost output—in both rich and poor economies.[22]

# 12

# Shock Cascades

*This time, like all times, is a very good one, if we but know what to do with it.* Ralph Waldo Emerson

THE INTERLOCKED SETS OF WORLDVIEWS, institutions, and technologies (our WITs) that undergird today's commitment to economic growth will change—and change radically.

But history tells us that when dominant WITs are replaced by new ones, the process is rarely smooth and peaceful. It usually involves a revolutionary shift in the social system—what I earlier called a "social earthquake"—and it's often accompanied by violence and trauma. Think of the rise of the European state system in the early modern period (1500 to 1800 CE), which was the product of the bloody Wars of Religion that extended across Europe over five generations; or of the overthrow of the worldviews, institutions, and technologies that supported slavery in the United States, which climaxed in the Civil War; or of the titanic struggle between Soviet communism and democratic capitalism through much of the twentieth century.

In the last case, democratic capitalism won, but the contest killed tens of millions of people in proxy violence between the two systems—from El Salvador and Angola to Afghanistan. Still, we should keep in mind, and be heartened by, the extraordinary end of this process: the popular uprisings against communism in Eastern Europe and the subsequent collapse of the Soviet Union itself were largely peaceful. And, like the democratic transition in South Africa at about the same time, this relatively benign outcome was almost entirely unexpected by experts.

Today, social earthquakes are happening far more quickly, and their effects can spread faster and farther than has been true historically, not surprisingly given the greater connectivity and uniformity within and among our societies. I call these factors "multipliers," because they can dramatically multiply the impact of other stresses on our societies. Greater connectivity and uniformity are both slow-process changes that have resulted from amazingly successful solutions to our everyday communication and transportation problems. And once more, in solving these visible and immediate problems, we've created some less visible, less pressing, and less tractable ones—in particular, a rising vulnerability to what complexity scientists call "cascading" failure. This greater vulnerability, however, may also create for us enormous opportunities for positive change.

## CONNECTIVITY AND UNIFORMITY

We all know that compared to, say, fifty years ago, connectivity between people, groups, technologies, organizations, and companies around the world has soared not just through flows of information—phone calls, text messages, torrent downloads, financial transfers, and the like—but also through flows of materials, energy, goods, and people themselves.[1] This extraordinary connectivity is now a defining feature of the tightly integrated social and ecological system we've created the world over. And even though populist nationalism has recently eroded political support in many countries for immigration and unrestricted trade, just about everything in our world—whether people, manufactured goods, or gigabytes—is still moving much faster, farther, and in vastly greater quantities than it did a half-century ago.

Also, in large part because of the shared commitment of the world's policy elites to economic growth, our institutions and technologies as well as some key components of our worldviews have become far more uniform. Nearly everywhere in the world, for example, national economic systems, including central banks, ministries of finance, and the like, have similar designs. Financial instruments like types of stocks, bonds, and their derivatives are similar nearly everywhere, too, as are computer hardware and software, social media platforms, antibiotics, core industrial processes, germplasm for essential crops and livestock (including humanity's 22 billion chickens, 1.2 billion sheep, and 1.5 billion cattle), fast-food restaurants, clothing, dominant political ideologies, languages of commerce, consumerist notions of the good life, and even blockbuster movies.

These two trends—higher connectivity and higher uniformity within our global socio-ecological system—reinforce each other in a feedback loop. In one direction, greater uniformity leads to more connectivity, because people link more easily with people they perceive as like-minded, organizations more easily with similar organizations, and technologies more easily with similar technologies. In the other direction, greater connectivity often leads to greater uniformity. Scientists have shown that as people, groups, and companies make new connections within their social or economic networks, they prefer to connect with other people, groups, and companies that are already highly connected. So existing patterns of connectivity are further reinforced, creating conditions for already dominant people, groups, and companies to become even more dominant—and the ideas they espouse more widespread.[2]

These processes make our societies' prevailing sets of worldviews, institutions, and technologies even more resistant to change, because we're obliged to opt into them if we want to gain the benefits they provide. For instance, to earn a good income in a capitalist market economy, we each need to sell our labor, skills, or ideas in that market, which means we must each accept the idea, at least for economic and employment purposes, that we're little more than commodities. And to the extent we all accept this idea—this "social fact"—we help reproduce the capitalist market WIT. We may not see ourselves as creators of the machine we live within, but we are.[3]

## CONTAGION 2008

The techno-optimists generally declare that more connectivity always makes us better off, and sometimes it's truly a good thing. As we saw earlier, modern global connectivity gives us the potential to supercharge combinatorial innovation: it helps the world's scientists, engineers, governments, firms, and citizen groups link existing ideas, institutions, technologies, and practices in new ways to create solutions to critical problems.

But while high connectivity might boost innovation, high uniformity often doesn't, because it can lower the likelihood of new combinations. Think of an immense stretch of farmland growing the same strain of corn in the neat, close rows that characterize modern industrial agriculture. The farmland's great expanse makes it a large system; the proximity of corn stalks to each other is analogous to high connectivity; and the fact that each corn plant has the same genes creates high uniformity.

These three features raise the farm's efficiency in good times: it can produce more tons of corn per dollar invested than can a smaller farm growing a diversity of corn varieties with plants spaced more widely. But the plants' genetic uniformity means there's little chance that new corn varieties will develop spontaneously in the field as the plants fertilize each other—varieties that might help the corn evolve to survive in, for example, a much hotter climate.

Researchers have found, too, that high connectivity across large systems can actually be dangerous when it's combined with high uniformity: together the two factors can make systems far more susceptible to cascading failure if hit by certain kinds of shock.[4] In the cornfield, for instance, high connectivity and uniformity together make the crop more vulnerable to sudden infestations of pests and diseases: if a blight affects plants in one small part of the field, the entire crop can quickly succumb as the disease jumps from one identical stalk to another.

This kind of cascading failure is an example of *contagion*. More connectivity enables change in one element to more easily cause change in another, so it's easier for the pathogen to jump between elements; and greater uniformity among the elements makes the pathogen's effects more consistently harmful.

These processes have, of course, already worsened the global spread of actual pathogens such as influenza, SARS in 2002, and COVID-19 in 2020. The human population is now among the largest bodies of genetically identical, multicellular biomass on Earth; taken together we weigh nearly a third of a billion tons. Combined with the connectivity arising from our proximity in huge cities and our constant travel back and forth around the globe, we're like an enormous petri dish primed with nutrients for cultivating and spreading new diseases.

But greater connectivity and uniformity have also already allowed economic shocks to propagate much more rapidly than they would have in earlier years—through global financial markets, food-trading systems, computer networks, and the like—making these shocks much more harmful.[5] The 2008–09 economic crisis was a case in point: stresses, in the form of sour loans, had been accumulating in the sub-prime mortgage market in the United States, a relatively minor part of the world economy, since 2006; but when the financial firm Lehman Brothers failed in September 2008, that shock abruptly cascaded into a global crisis. In a matter of weeks, world financial markets seized up, liquidity vanished, and international trade collapsed.

The calamity was both global and virtually instantaneous. It hit every major economic region in the world simultaneously, including China, because of extreme connectivity arising largely from highly interdependent balance sheets (firms owning a lot of each other's shares or bonds, for instance), highly uniform financial instruments, and trading links tightened by just-in-time supply chains. In contrast, the Asian financial crisis in 1997–98 and the dot.com implosion in the early 2000s were both largely regional; and the effects of the financial crash that precipitated the Great Depression in 1929—an event with which the 2008 panic is often compared—took months to spread fully around the world.

As credit dried up from commercial banks, governments and central banks injected trillions of dollars of liquidity into the global economy to try to keep businesses and consumers buying. Yet a bitter recession still followed, focused in Europe and North America. People and communities were devastated. In the decade after, in economies where growth eventually returned, gains went mainly to those in the top few percentiles of wealth and income. Vast numbers of others, particularly young people, in countries on every continent remained mired in unemployment and underemployment, stagnating or falling incomes, narrowed work opportunities, and chronic economic insecurity.

The crisis, it turned out, marked an abrupt nonlinear shift in the complex world economy—a flip from a state of relatively high growth and modest inflation to a state of more modest growth flirting with deflation. Whereas between the mid-1980s and 2008, world growth rates, outside of recessions, fluctuated between roughly 3 and 4.5 percent annually (averaging about 3.8 percent), from 2011 to 2019 they stayed between about 2 and 3.1 percent (averaging about 2.8 percent)—or 1 percent lower than before, which is a large difference when compounded over time.[6] This change, I believe, indicates that the longer-term trend towards lower economic growth—anticipated by economist Robert Gordon—is now underway, though the effects on this trend of the 2020 economic downturn due to the pandemic remain to be seen.

The financial crisis was actually only one of three social earthquakes that shook the world simultaneously in 2008. Between January and June of that year, as the US sub-prime mortgage crisis was reaching its climax, world energy prices soared—the international price of light crude oil rose more than 60 percent to over $140 a barrel—and the price of grain worldwide shot upwards too, triggering food riots and violence in dozens of poor countries. Few commentators or analysts have noted the extraordinary synchronicity of these three crises; but they were intimately related to each other.[7]

### OVERLOAD

Cascading system failures can also arise from *overload.* We're all familiar with the basic concept. We know we're likely to make mistakes, even freeze up, when we try to pay attention to too many things at the same time, especially when they're novel and unexpected. Think of driving down a busy, unfamiliar city street at night, watching the GPS map on the dash, with streetlights changing, pedestrians jaywalking, advertisements flashing on every side, and kids yelling at each other in the back seat. Our brain—specifically its executive functions mediated by the prefrontal cortex—can become overwhelmed with information and stress, so our ability to make complex, accurate, and effective decisions plummets. And then we're much more likely to have an accident.

Engineers have long known that overload can cause failures in the highly connected networks that carry our societies' information, energy, and materials.[8] When a link between nodes in one of these networks fails, the flow of stuff along the link is displaced onto other, nearby links. That extra flow can then overload those links and make them fail too, leading to a displacement of even larger flows onto yet other links that fail in succession. This was the mechanism behind the huge electricity blackout in North America in August 2003: a single high-tension line shorted out in rural Ohio, triggering grid failures that quickly affected over fifty million people across nine states and provinces.

The process can operate in our highly connected social systems, too. The world's rising stresses, synchronized crises, and torrents of information now often overload the links and mechanisms—the channels of communication, the day-to-day rules and procedures, and the political and economic institutions—that our societies use to stabilize and manage their affairs. Decision makers dash madly from one crisis to another, and institutions buckle under the load of demands and expectations as they fall further and further behind in the race to deliver even marginally effective solutions to the problems they're supposed to address. Then, powerful groups and opportunistic politicians exploit the rising dysfunction—and the anger the public feels over manifest government incompetence—to find openings to advance their narrow interests.

Take the European Union's reaction to the tide of war refugees and migrants pouring over its borders in 2015 and 2016, many fleeing the Syrian civil war. The sheer number of arrivals swamped Europe's standard procedures for managing immigrants, and a coordinated European response collapsed. Then, countries along Europe's southeastern boundary broke ranks and enforced their own crisis policies, some erecting hundreds of kilometers of barbed-wire fences. The result: tens of thousands of destitute refugees were stranded in tent-city squalor along Europe's edge, while populist politicians across the continent used the chaos to whip up xenophobic emotions and attack the European Union.

### LOCK IN AND BREAK OUT

It's easy to see now how slow processes can generate fast processes: initially, pressure from tectonic stresses builds up within rigid social systems; then, when this pressure becomes extreme, a seemingly small event triggers the pressure's release as a social earthquake.

In such situations, higher connectivity and uniformity can combine to produce two key effects—one that reinforces rigidity and another that, somewhat paradoxically, magnifies change once it starts to happen.

At first, connectivity and uniformity reinforce rigidity by encouraging people to opt into already dominant sets of worldviews, institutions, and technologies (WITs), a process that helps "lock in" our societies' current structures of social power and keeps them from making the necessary changes to accommodate the rising underlying stresses. But once the stresses have built up to the point where a sudden shift occurs, connectivity and uniformity then amplify the speed and extent of social earthquakes, as dominant WITs begin to be replaced and structures of power start to break down.

We usually assume that such social earthquakes are deeply harmful. And honest hope demands we be clear: today's children, and even most adults alive today, will see, during their lives, pulses of harmful social earthquakes, including economic crises, pandemics, fights over scarce resources like water, food price shocks, torrents of refugees and migrants crossing borders, and widespread civil violence, terrorism, and war.

But we can acknowledge these likely prospects without capitulating to them. And if we want our hope to be astute and powerful, too, which means strategically smart, we need to understand the latent potential in our world's social systems for other, better possibilities—and we need to figure out how to exploit that potential.

The examples of the largely peaceful collapses of apartheid and the Soviet Union—and even of the wide political mobilization that led to the global treaty banning atmospheric testing of nuclear weapons, which Stephanie May and so many others struggled so hard to achieve—tell us that beneficial outcomes, while perhaps rare, are still very much possible. The breakdown of dominant structures of power can sometimes lead to "break out"—a rapid, radical, but still mainly humane escape from the entrenched social reality that dominant worldviews, institutions, and technologies create. And this is exactly what the young climate activists who have mobilized around the planet are hoping to achieve.

## 13

# A Message from Middle-earth

*Until you start focusing on what needs to be done rather than what is politically possible, there is no hope.* Greta Thunberg

BEING A PARENT OF YOUNG CHILDREN, I've happily discovered, means getting a second chance to do things I missed in my own childhood—at least the things that won't leave my body in splints and braces. I've had particular fun spending hours reading children's classics aloud to my kids that I'd skipped when I was young or that hadn't been written yet—Lucy Maud Montgomery's *Anne of Green Gables*, Kenneth Oppel's *Airborn* series, Philip Pullman's *Dark Materials* trilogy, and of course all seven volumes of *Harry Potter*—twice.

One tome that I'd never mastered was J.R.R. Tolkien's *The Lord of the Rings*—science fiction interested me far more than epic fantasy and myth. So when I opened volume one, *The Fellowship of the Ring*, on an icy December evening in 2013, cuddled with then eight-year-old Ben in an armchair, I wasn't exactly a study in enthusiasm.

Tolkien isn't for everyone, that's for sure. *The Lord of the Rings* is a prototypical hero's quest that's full of daring adventures, countless bloody battles, and much valor to protect honor, friends, tribe, and the truth. With only a couple of notable exceptions, powerful, courageous, and clever women don't cross the pages. And even taking into consideration the genre's prevailing style, the prose is—how can one be polite?—a tad labored.

Yet I was soon enthralled. The saga captivated me in some primal way, as it has so many others. And by somewhere in the middle of the first volume, I realized why I was hooked. *The Lord of the Rings* is an extended meditation on what one should do when things appear utterly hopeless. It is, in fact, an account of how to survive by creating and living through hope that's astute and powerful. Such hope is informed by a deep understanding of the minds of people we encounter—friends, allies, and enemies—as we strive to reach our vision of a positive future. And it employs that knowledge in ways that are strategically smart.

If you're not familiar with the story, here's the gist. And bear with the recap—it is relevant! The Dark Lord Sauron has established himself in Mordor, a forbidding, barren, and mountain-ringed domain in the eastern reaches of Middle-earth. From there, he's sent forth his brutal servants to find and bring to him the One Ring, which will give him unassailable power over all lands. But that ring is currently in the possession of Frodo, a middle-aged member of the race of Hobbits—short, peaceable, human-like creatures with very hairy feet (to Ben's delight). The wizard Gandalf tells Frodo that the only way to keep the ring from Sauron forever is to take it into the heart of Mordor itself and throw it into the volcanic fires of Mount Doom. That idea seems patently absurd. Not only does the ring corrupt everyone who touches it (although it seems to have less effect on Frodo), Mordor seethes with teeming masses of vicious creatures called Orcs, spectral beings on flying dragons, and various other nightmares, all overseen by Sauron's panoptical eye atop a black tower that reaches far into the sky.

To make matters worse, the main races that populate Middle-earth outside of Mordor—Hobbits, Elves, Dwarves, and Men—have squabbled among themselves for centuries. Even if they could somehow manage to pull themselves together, their collective armies aren't strong enough to penetrate Mordor's mountainous defenses to reach Mount Doom, while the idea of sneaking past Sauron's vast forces and under his watchful eye seems preposterous. The situation is surely hopeless.

But it's not, we learn. Getting the ring to Mount Doom and defeating Sauron involves a healthy portion of luck, a good quantity of magic (the tale is, after all, a fantasy), but also—though much less remarked upon by readers and literary critics—an enormous amount of emotional, interpersonal, intercommunal, and strategic smarts.

Early on, a team is formed—the Fellowship—to solve the problem. It's a cantankerous bunch that often comes close to breaking apart. But it incorporates at least one representative from every major race Sauron threatens. Over time, and under extreme duress, they learn to respect and even love one another. Equally important, though, is the mix of talents the team possesses: each member contributes something vital and unique to the enterprise's ultimate success.

At first, none of them has a clue how to get the ring to Mount Doom: the further they peer into the future the worse the uncertainties become. As Elrond, the Half-elven Lord of Rivendell, advises: "None can foretell what will come to pass, if we take this road or that."[1] So they keep their initial plan simple and start by exploring the adjacent possible—that zone of uncertainty just beyond the edge of the known—by taking one step at a time closer to Mordor. They recognize they must figure things out as they go, constantly recalibrate their plans as they learn more, and constantly manage their doubts and fear. Along the way they wonder if they should turn back; divert themselves to lesser, more achievable goals; or split apart and return to their homes to defend their respective racial groups against the coming storm. What little hope they have is under incessant siege.

The Fellowship is soon sundered by deceit within and attacks from without. Later, some members reunite and then split apart again, while Frodo and his friend Sam seek to reach their destination alone. At least they think they're alone. Actually, though, every other member of the team repeatedly deliberates on how to help Frodo and—even in the absence of communication among them—acts on those deliberations. The story's climax arrives when most of the team, finally reunited, take much of what remains of their armies to a distant boundary of Mordor to distract Sauron, so Frodo and Sam can climb to the top of Mount Doom unobserved. Beforehand, the team debates at length the ridiculously low odds of their plan's success. They know it's a huge gamble—the equivalent in modern sporting parlance of a "Hail Mary" pass—that depends on their estimates of Frodo's physical and mental state and of his location on the slopes of Mount Doom being exactly right.

No surprise, their estimates are right, the One Ring is destroyed, albeit somewhat by accident after Frodo yields to its powers, and Middle-earth is more or less saved, although profoundly and irreversibly changed. Good wins, sort of, and evil is vanquished, at least for the time being, as I learned when I turned the third volume's last pages eight months later, snug in a tent with my family in the pouring rain during an August camping trip.

### AMDIR AND ESTEL

A deep dive into Tolkien's classic fantasy might seem a little out of place in a sober book about hope. But his story is not timeless without reason. It shows that for our hope to be both astute and powerful, we need a clear, motivating vision of where we want to go, a way of identifying which paths to take, and a thoughtful understanding of the worldviews and motives of the people who can help us along the way—and of those who might try to stop us. It also reminds us that we'll be far more effective if we're ready to recalibrate our vision, our judgment of the best paths forward, and our assessments of others' views as we go along.

Hope is unquestionably one of Tolkien's central concerns in *The Lord of the Rings*; the word itself appears about three hundred times in the book's three volumes. Tolkien, a scholar with vast knowledge of history and myth as well as a keen observer of human nature and our moral struggles, had been deeply marked by World War I, having fought as a young officer in the Battle of the Somme and having lost almost all his male friends to the war. Also, he wrote *The Lord of the Rings* in a time and place—the England of World War II, 1937 to 1949—when hope was under siege and eventually rescued through almost unimaginable courage, toil, and sacrifice, and with the help of many allies. We tend to forget how bleak the situation must have looked to Tolkien and his compatriots in the early days of World War II, when "little England," isolated on the edge of Europe and appallingly ill-equipped, stood almost alone against Hitler's staggering forces.

Tolkien later dismissed attempts to draw allegorical parallels between those events and his story. Yet they may have helped him arrive at his sophisticated and subtle understanding of hope, which was revealed in unpublished passages edited and released posthumously by his son. In a conversation about hope that Tolkien crafted between an Elf king and a wise woman, he distinguished two kinds: "Amdir," a hope that involves "an expectation of good, which though uncertain has some foundation in what is known"; and "Estel," a deep hope born of trust or faith that things will turn out well. When Amdir succumbs because we see no escape, Estel can remain steadfast.[2]

Many Christian commentators and scholars, wanting to claim Tolkien's mantle (he was a devout Roman Catholic), say he espoused a Christian hope based on faith in God's ultimate intervention and redemption. By this view, hope, which in this case would be Estel, can remain secure, because we know God will take care of us in the end, even if all evidence around us says we're doomed. Other Tolkien aficionados vehemently argue that he eschewed hope entirely, regarding it as the sentiment of fools; his protagonists keep going because of nothing more than their ardent commitment to courage and cheer, regardless of what the future seems to hold.

Neither argument convinces me. There's little hint of Christian eschatology in the pages of *The Lord of the Rings*; and the book's life philosophy is deeply informed by Norse, Germanic, and Celtic myth. Indeed, to my mind, Tolkien's heroes possess the Finnish virtue *sisu*, which translates roughly as "fierce tenacity" or "toughness," and indicates inner moral strength in the face of daunting odds. Tolkien's protagonists regularly express, as well, something akin to Amdir, a hope grounded in evidence and reason. They might think that the chance of a good outcome is terribly slim, but they pursue that chance all the same, with eyes fully open to the risks such a choice entails.

But mostly I'm not convinced because in the book Tolkien argues time and time again for another source of hope—the same source, it turns out, I mentioned that book-tour evening when asked "Is it hopeless now?" Tolkien identifies this source most clearly in Gandalf's remarkable words to the Council of Elrond, when the Fellowship is formed. Pressed on whether the effort to take the ring to Mount Doom is a path of despair or folly, the wizard replies: "Despair or folly? It is not despair, for despair is only for those who see the end beyond all doubt. We do not. It is wisdom to recognize necessity, when all other courses have been weighed, though as folly it may appear to those who cling to false hope."[3]

Honest hope, Tolkien implies here, echoing Reinhold Niebuhr's contemporaneous Serenity Prayer that I quoted earlier, emerges from the wisdom that recognizes necessity. Think of the necessity imposed by the ineluctable slow-process constraints on future possibility that we looked at in chapter 9. Critically, though, within the boundaries imposed by that necessity, we can't know for sure what will transpire—we can't "see the end beyond all doubt." And it's there, within that deep uncertainty, Tolkien seems to be saying, that we can find Estel, a hope that trusts in the world's complexities to produce as yet unseen possibilities, some of which might be good.

Tolkien seems to suggest that people keep going in dreadful circumstances through a combination of, or sometimes an alternation between, Amdir and Estel. Amdir is hope with an object—"an expectation of good" arising from a vision of a positive future. But whether this vision will ultimately be realized is highly uncertain, given "what is known." Estel, in contrast, doesn't have an object; the solace it provides arises instead from its confidence in uncertainty's promise, all the while acknowledging that this promise may not, ultimately, be realized in something good.

Throughout *The Lord of the Rings*, Tolkien also implicitly asserts that hope (in this case, Amdir) should be smart. For our hope to be smart, we need a strategy for quickly and effectively searching the adjacent possible to identify—and provide rough-and-ready evaluations of—possible paths forward. When the range of paths is immense, as it was for the Fellowship, the findings of complexity science indicate that the search will often work best when undertaken by a highly diverse group of agents (like the Fellowship's members), linked together in a loose non-hierarchical network (so that each agent has a chance to discover new possibilities and then share news with everyone else), guided by only rough goals, and open to constant reformulation of its plans. This is, in fact, how complex systems as diverse as competitive economic markets and the immune systems of mammals solve their problems—it's usually the fastest and most efficient way to find good possibilities amidst a welter of noise, distractions, and false leads in the marketplace, or in our own bodies.[4]

Tolkien seems to have grasped this reality. But he then gave a special twist to the idea: we should often look for good possibilities, he advises, *in the opposite direction* of what seems to make most sense at the time. So, rather than fleeing from Sauron's massed forces in Mordor, which common sense would seem to counsel, perhaps the Fellowship should go straight towards them. And rather than attacking Sauron directly with all the armies the Fellowship can muster, perhaps the Fellowship should send in only one or two of the least formidable beings in the land. Seek solutions in unexpected places, in other words.

Finally, Tolkien shows that hope needs to be emotionally and psychologically smart too. The story's protagonists discuss at length each other's characters, perspectives, and intentions, just as they discuss those of their potential allies and likely enemies. They anchor their decisions in a keen awareness of what's going on in their own heads—of their own beliefs, values, and deep motivations—and, as much as possible, of what they think is going on in the heads of the people and groups who might ally with or oppose them.

### AN UNFORGIVING DILEMMA: ENOUGH VS. FEASIBLE

Like Tolkien's Fellowship, today's children and adults face a landscape of possible paths into the future that is, frankly, terrifying in its complexities and risks. If we're going to choose a good path—good not just for ourselves narrowly, but for all of humanity—we need discriminating criteria for selecting among the multitude of options we face.

Here are two criteria, which I mentioned briefly in chapter 2: first, any path we choose should take us towards a future that's likely, as best we can judge, to be *enough* to genuinely reduce the danger humanity faces; and, second, the path should also be *feasible*, in the sense that it gives us a good chance of reaching that desirable future, which means surmounting or bypassing likely political, economic, social, or technological roadblocks along the way. These criteria are often at odds, leaving us trapped in what I call the *enough vs. feasible dilemma*.

People who care about humanity's future—and want to do something to help—generally cope with the dilemma in one of two ways. Some who are idealistic—some activists in civic groups and many academics—focus on making sure their proposals for solving humanity's problems would be enough. They largely ignore whether the proposals are feasible, implicitly downplaying political, economic, social, and technological obstacles to their implementation. They hope that either those obstacles will turn out to be far less severe than seems likely or that a new social phenomenon—a *deus ex machina*, like a revolutionary worldwide social movement—will sweep them aside. Basically they say: We should start now to do *whatever's necessary* to solve the problems, and deal with obstacles to implementing our solutions later, if it turns out that we must.

Others—for example politicians, policy makers, corporate leaders, and researchers in think tanks who see themselves as anchored in the practical world—focus on making their proposals feasible and largely ignore whether they'd be enough. They tend to highlight the obstacles to implementing solutions, implicitly hoping that the problems themselves will turn out to be far less severe than evidence today indicates is the case—or easier to solve sometime in the future, maybe with a technological breakthrough. Basically, they say: We should start now to do *whatever's possible* to solve the problems, and deal with whatever remains of those problems later, if it turns out we must.

Both groups, in other words, cope with the dilemma by implicitly ignoring or wishing away the side—or "horn"—of the dilemma they don't like, despite the evidence. They either posit changes in our societies and technologies that seem far-fetched (a kind of magical thinking), or they deemphasize the seriousness of humanity's problems (a kind of denial). Each group lives in its own la-la land, while blaming the other for wasting time and resources on fantasies or for not doing enough. And as humanity's situation worsens, the gulfs between these la-la lands and our increasingly troubling reality are becoming positively pathological.[5]

Neither approach is far from the "How about" imaginings of kids, which could be good, except that they both neglect the tempering realism that ideally comes with adulthood.

## SUPERORDINATE GOAL

The protagonists in *The Lord of the Rings* face a version of this dilemma. The single action that would be enough to stop Sauron forever—throwing the ring into Mount Doom's fires—doesn't seem feasible, at least on first assessment. Yet other actions that are feasible—trying to hide the ring in the countryside, giving it to the ancient creature Tom Bombadil, or dropping it into the deep sea—ultimately won't stop Sauron.

The Fellowship's dilemma seems impossibly difficult, but it's actually more tractable than humanity's version today. First and most obviously, Tolkien creates for his group of heroes a conventional "war" problem: an external and willfully malicious enemy threatens the group with annihilation, which catalyzes its members to bury their differences and collaborate to defeat the enemy.

In contrast, today the enemy isn't "out there" but at least partly "in here," right inside our own societies, communities, families, and even ourselves individually. Our severe problems are rooted in factors like bad technologies, poorly designed institutions, greedy corporate elites, aggrandizing states, self-interested consumers, and ingrained patriarchy and racism. And they're rooted in deep aspects of our human nature, like our rapacious appetites as top predators, our tendency to prefer our own group to other groups, our powerful desire to procreate, and, not least, our seemingly boundless capacity to lie to ourselves when faced with unpleasant truths. As in the sinking lifeboat, nearly all of us have contributed somehow—if to greatly unequal degrees—to the dangers we're facing. To quote the wonderful mid-twentieth-century American political cartoonist Walt Kelly, "We have met the enemy, and he is us."[6]

But in *The Lord of the Rings*, Sauron isn't just an external, personified enemy; he also creates for the Fellowship what social psychologists call a "superordinate" goal—one that's clear and specific, that everyone shares, that overrides all others, and that can't be achieved without cooperation.

These features encourage teamwork among the members, and this teamwork makes reaching the goal—to throw the ring into the fires—far more feasible, which ultimately helps the Fellowship solve their enough vs. feasible dilemma. (In the classic demonstration of the power of superordinate goals, the Robbers Cave Experiment run by social psychologists Muzafer and Carolyn Sherif in the mid-1950s, antagonism between two groups of boys at a summer camp was sharply reduced when, among other collaborative tasks, they had to pull on the same rope together to get their food truck restarted.[7])

Today, though, humanity has a profusion of big problems. Worse, because of both deep uncertainty about the future and bitter ideological differences among us, we heatedly disagree about what our problems are and their severity, about whose responsibility they are and to what extent, about what their solutions should be, and about whether those solutions will work, even if we could implement them. Alas, compared to the Fellowship's situation, ours is a massive muddle. Instead of drawing us together, this muddle often just splits us further apart, which only further erodes our ability to find effective and feasible solutions and, in the end, erodes our hope too. Tolkien, for all his understanding and smarts, can't take us all the way.

## SWEET SPOT

Climate change, for example, has emerged as a potent threat to our well-being as a species, especially to that of our children and grandchildren. International scientific bodies have accepted its reality for decades, and nearly all national governments now accept it, too, apart from those of several large fossil-fuel producers, including Russia, Saudi Arabia, and the United States under the Trump administration. Indeed, climate change is quite possibly an existential threat to all our societies. So solving this problem should be a potent superordinate goal for humanity. And that's exactly how many world leaders, such as all recent United Nations secretaries-general, as well as international agreements like the 2015 Paris Climate Agreement, have characterized the challenge.

Yet the data show that most people around the world don't see the challenge this way, not so far. True, they're worried, and many even see climate change as a preeminent danger. The Pew Research Center reported in early 2019 that 67 percent of people polled in a worldwide survey believe climate change is a "major threat to our country" (up from 56 percent in 2013), ahead of cyberattacks, Islamic State terrorism, North Korea's nuclear program, and problems with the global economy.[8] It was ranked as a top global threat, often by large majorities, not only in many countries in Europe and North America, but in countries across the Asia-Pacific region, Africa, Latin America, and the Middle East.

Yet the problem is still identified as a threat to one's own country, not as an existential threat to humanity as a whole—a threat that overrides all others and demands species-wide cooperation. And this is the main reason why, when it comes to climate change, we're firmly stuck on the horns of a wicked enough vs. feasible dilemma. Responses that could be enough don't seem remotely feasible, and those that could be feasible won't be remotely enough.

What are these responses? They fall into three broad categories: adaptation, mitigation, and geoengineering.

*Adaptation* accepts that at least some warming is going to happen and aims to minimize that warming's harm—by doing things like building bigger storm sewers and sea walls and moving people away from zones vulnerable to drought, heat waves, and floods. *Mitigation* means cutting greenhouse gas emissions by, for example, taxing carbon and replacing fossil-fuel energy with solar, wind, and other renewables. *Geoengineering* tries to deal with the problem after the fact by changing entire planetary systems; possibilities include injecting sulfur dioxide crystals into the high atmosphere to reflect sunlight back into space or sucking colossal amounts of carbon dioxide out of the air and then pumping it underground.

People in countries around the world and across the ideological spectrum largely agree that all societies will need to adapt—at the very least. Yet, even though warming is already having large and sometimes devastating impacts just about everywhere, no country—none—has embarked on the deliberate and coordinated adaptation response that's increasingly needed. Adaptation has been implemented in bits and pieces, sometimes within cities or at the level of provinces or states; and some corporations, like insurance companies, have started to adjust their plans (by, for example, refusing to cover certain flooding threats). But when it comes to whole countries, all have dropped the adaptation ball. And very few, except for a handful in Europe like Denmark and Germany, have undertaken truly meaningful efforts to cut emissions either; almost all national mitigation programs only barely nibble at the edges of the problem.

These shortcomings are widely recognized. Yet few people realize that through the 2015 Paris Agreement, the world's countries also implicitly committed themselves to a form of geoengineering—in this case the extraction of carbon dioxide from Earth's atmosphere.

Andrei Popov

But that's true. The agreement's plan to cap warming at no more than 2 degrees Celsius, and ideally limit it to 1.5 degrees, is based on computer projections that, in almost every case, require humanity to take immense quantities of carbon from the air and put it somewhere it can't do any harm. Technologies barely imagined today will be needed to generate these huge "negative emissions." One possibility is that our children will plant fast-growing trees across vast territories to absorb carbon dioxide as they grow; then they'll cut the trees down to produce energy and pump the resulting carbon dioxide underground.

This might be *enough* to halt climate change and perhaps even reverse it; but it doesn't currently seem feasible. To reach the agreement's goals, the scale of the proposed projects would have to be breathtaking. Keeping temperatures from rising above the 1.5-degree target in 2100, for instance, would require removal from the atmosphere of at least a half-trillion tonnes of carbon dioxide, in a global effort starting almost immediately and extending beyond the end of this century.[9] That amount would fill sixty-five Grand Canyons or a balloon filled with pure $CO_2$ measuring about eighty kilometers in diameter (at sea-level atmospheric pressure), and removing it would entail the largest industrial project in history by far, one that would absorb a large fraction of the world's economic output for decades.[10]

Implementing this project on the scale required would be monumentally costly therefore, and it would demand maybe a century or more of broad and deep cooperation among the world's nations, probably involving some kind of global government that could coerce recalcitrant countries—a prospect that today seems remote.

On the other hand, some kinds of adaptation to climate change are certainly *feasible*, in the sense that there's no doubt that, if we try, we can install bigger storm sewers, build gigantic sea walls, and the like. But these efforts won't be enough, because while they might reduce the danger climate change poses to some of us for a time, they won't do anything to address the problem's underlying causes—vast and growing emissions of carbon dioxide and other greenhouse gases, which until the economic downturn of 2020 were rising between 2 and 3 percent a year and which will likely resume rising as the global economy recovers. So if we pursue adaptation alone, climate-related disasters will become more frequent and severe—and will eventually overwhelm all our efforts to adapt.

That adaptation won't be enough by itself isn't widely grasped. Yet without greenhouse gas emission cuts, warming will likely surpass 4 degrees sometime in the second half of this century, according to recent research, depopulating large areas of the globe through heat stress and causing widespread societal breakdown in the remaining regions, mainly because of agricultural failure. The idea put forward by some techno-optimists, like Bill Gates, that we can invent our way out of a food crisis by developing genetically modified crops adapted to such a world, is sheer idiocy. At 4 degrees Celsius, British climate scientist Rachel Warren notes, "the limits for human adaptation are likely to be exceeded in many parts of the world, while the limits for adaptation for natural systems would largely be exceeded throughout the world."[11]

What, then, about mitigation? A policy of drastically reducing greenhouse gas emissions once seemed both enough and feasible. But as time has passed and efforts to implement major cuts have repeatedly failed, many experts have concluded, to their despair, that truly meaningful emission cuts simply aren't going to happen in the foreseeable future, because they aren't politically, socially, and economically feasible, especially given the implacable opposition of fossil-fuel vested interests and their political allies. And, now, because humanity has already discharged massive amounts of carbon into the atmosphere, cuts large and fast enough to hold warming to 1.5 degrees will be virtually impossible to accomplish. By 2030, according to the United Nations, humanity must reduce global output of greenhouse gases by 60 percent below projected levels just to preserve a two-in-three chance of keeping warming under 1.5 degrees by 2100. In other words, in just over a decade humanity must shave thirty-five billion tonnes each year off its projected greenhouse gas output of nearly 60 billion tonnes.[12]

That's not going to happen. Despite the optimistic language from international climate negotiations, humanity has waited too long to get serious about mitigation, so warming beyond the 1.5-degree threshold is now inevitable. It will be one of the slow-process structural factors profoundly shaping humanity's future possibilities.

I've created some simple diagrams that I find helpful in clarifying the enough vs. feasible dilemma we face with the climate crisis.

In the first diagram, the outside oval represents the full set of humanity's potential responses to the problem—all the things we might do in the most expansive sense to address climate change, even if some seem ludicrous. Inside this big oval you'll see a smaller one that stands for the subset of all responses that would be *enough* to keep warming to no more than, let's say, 1.5 degrees. Each response in this oval could, if implemented, very likely keep us within 1.5 degrees. An example would be an emergency program—coordinated and implemented globally—that, in line with the just-mentioned UN research, cuts world fossil-fuel use 60 percent below projected 2030 levels by deploying solar and wind farms around the planet.

The other oval in this first diagram represents the subset of *feasible* responses. Each response in this oval, if attempted with what we might agree is a reasonable effort, would have a good chance—say, one in five—of being implemented. An example would be a set of government regulations, like the vehicle fleet fuel-efficiency standards the Obama administration established in the United States

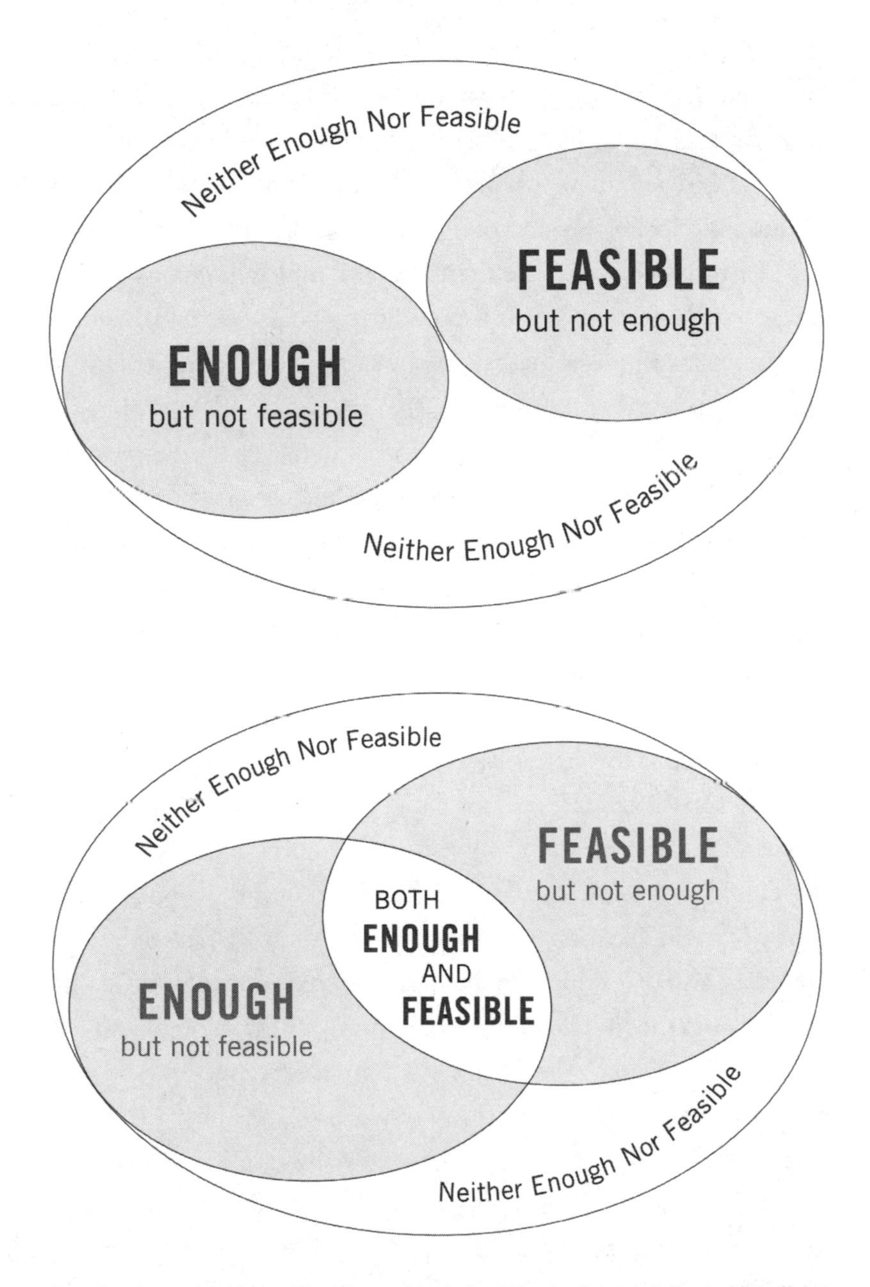


*A good response to a problem like climate change should be simultaneously "enough" (sufficient to genuinely reduce the danger the problem poses) and "feasible" (realistically achievable).*

(and that the Trump administration subsequently weakened), obliging automobile manufacturers to double the gas mileage of our standard car engines so they emit less carbon dioxide. Those responses outside the oval aren't feasible in this sense, because of some combination of technological, economic, social, or political obstacles.

The way I've drawn the first diagram, the enough and feasible subsets don't overlap. If this is our climate change reality, we must choose from responses that fall in one or the other category, or neither, because no responses are *both* enough and feasible. The dilemma is absolute, and halting climate change at 1.5 degrees isn't possible. But in the second diagram, the dilemma can be resolved, because some responses—those that fall in the sweet spot where the subsets overlap—are both enough and feasible. If this is our reality, to fix climate change—or at least to keep warming to 1.5 degrees—we need to find responses that fall into this zone.

But the history I've described of our attempts to reduce carbon emissions points to a larger lesson about why we've not found such responses to the climate crisis yet, and more generally why humanity hasn't pulled together to solve its critical problems. For most of these problems—not just climate change, but also collapsing biodiversity, rising economic inequality, and the constant danger of nuclear war—opposition by powerful vested interests makes the sweet spot in the middle of the diagram, where the enough and feasible subsets overlap, small and sometimes tiny, and maybe even nonexistent unless that opposition is overcome. Proposed solutions that might be enough run headlong into antagonism, sometimes vicious, from powerful groups—companies, business associations, trade unions, bureaucracies, citizens lobbies, and the like—fighting to preserve the status quo, which makes them not feasible.

So at best our societies, and all of us, in our efforts to address these problems, have been implementing small and largely non-disruptive changes to existing institutions, policies, and technologies—changes that are hardly ever enough.

### SOLVING THE DILEMMA

I've argued that any proposed response or responses to any one of humanity's grand challenges should pass a rough, two-part test: Would it or they be enough to make a real and positive difference to that challenge if implemented? And is implementing it or them feasible?

The few responses that pass both parts sit in the zone where the enough and feasible subsets overlap, and they can help us identify a good path for humanity into the future. This two-part test also helps keep our hope honest: if we're honest with ourselves, it doesn't make sense to invest hope in responses that can't be enough or won't be feasible, because then we're hoping for an outcome that's not desirable or pretty hopeless.

So far so good, but complications inevitably follow. Since we're uncertain about what kind of future we want and about the extent of our agency, notably our capacity to create a future we want, we don't really know how much the enough and feasible subsets overlap, or if they overlap at all.

Still, there are two general ways we can move these boundaries to expand the enough-feasible zone, or perhaps even create that zone if it doesn't already exist. First, we might try to enlarge the *feasible* subset. The easiest way is to simply lower the bar on what we mean by "feasible." If, for instance, we say a response to a problem counts as feasible if it has just a one-in-ten chance of implementation, rather than a one-in-five chance, we'll likely boost the number of responses we can say are feasible.

But that's just tinkering with definitions. A more substantial way of enlarging the feasible subset exploits our recursive imaginations.

We can travel in our minds into the future—into the adjacent possible—to create new combinations of what we know and to explore "How abouts" there. When we discover potential solutions in those imaginary futures that could be enough, we can then mentally experiment with ways of making those solutions feasible. This is how Tolkien's characters created the conditions to justify honest hope: they individually and collectively imagined and mentally experimented with possible paths to the ring's destruction, and then they chose one path and pursued it. Later I'll describe tools that can help our imagination in such experiments by, for instance, helping us frame our common problems as superordinate goals or see ways to better overcome vested interests that block solutions to these problems.

The second general way that some people might use to expand the enough-feasible zone is to try to enlarge the *enough* subset. Just as we can lower the bar on what counts as feasible, we can lower the bar on what counts as enough, but this time it involves more than just

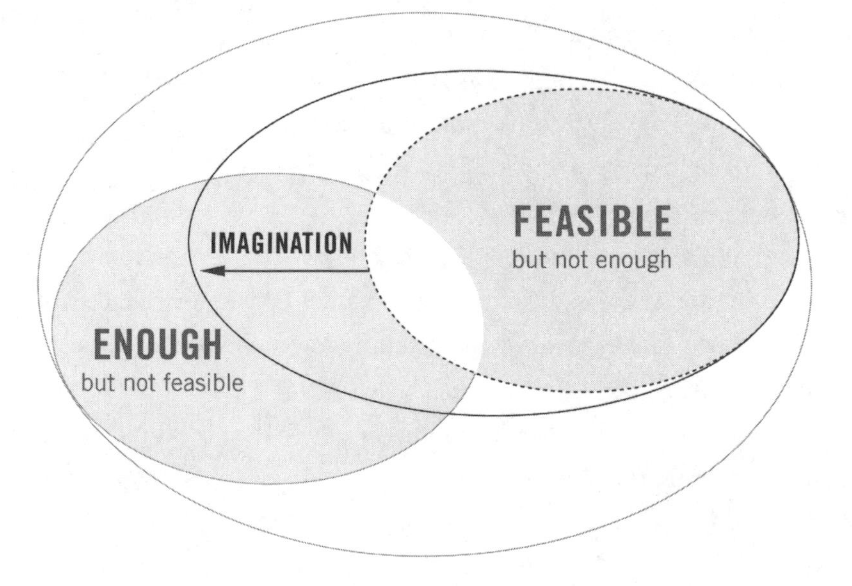


*We can use our imaginations to expand the set of responses that is feasible.*

tinkering with definitions. By lowering our standards of what counts as a satisfactory response to a problem, we can boost the number of responses in the enough subset. If, for instance, we're prepared to accept much greater future damage from climate change—so that we need to cap warming at only, say, 3 degrees, instead of 1.5 or 2 degrees—then a broader range of future energy systems could do the job (that is, could be enough to keep us under 3 degrees).

This example shows that what we count as "enough" as a response to any given problem—indeed, what we count as a problem itself—depends on our broader worldview, and particularly on the values, identities, and goals that are vital elements of our worldview. In other words, for some people, even for some who understand and accept the basic science of climate change, letting our warming target slip from 1.5 to 3 degrees isn't a big deal. They're willing to slice off a pound of flesh here or there; they don't care much about protecting the natural world or the well-being of people who'll be hit hardest by warming's impacts, such as poor people living in geographies vulnerable to extreme weather or today's children who'll endure climate crises later in their lives.

The object of our hope—our vision of a positive future that embodies our most fundamental moral commitments and that's often a key component of our worldview—strongly influences how we assess a given response to a problem like climate change. If we hope for a world where poor people and children are protected from warming's ravages, then we'll judge a response that only halts warming at 3 degrees to be not remotely enough.

### TRIAGE

As global crises worsen through the coming decades, humanity as a whole and each of us as individuals will often feel obliged to relax our goals—to compromise our hope by allowing our vision of the future to become dimmer and less positive. In fact, in an age of deepening loss, triage—the act of sorting things according to whether we think they can be saved or not—may become the principle that guides our survival.

For instance, as climate change advances we're already sacrificing species of animals and plants, often unwittingly. Through collective ignorance and inaction, we're deciding, basically, that some—like the staghorn corals of the Gulf of Mexico, the Arctic ringed seal, and the golden toad that used to live in the cloud forests of Central America—aren't important enough to save. By the middle of this century, humanity will likely be sacrificing whole geographic regions, consciously and deliberately at that point, as coastlines are abandoned to flooding and some tropical zones become too hot to inhabit. And given our history, it's not a stretch to imagine that towards 2100, powerful elites could be, in essence, sacrificing entire communities of people that, they've decided, are outside their "we" and so can't or shouldn't be saved.

In effect, when we put values, identities, and goals that justify triage at the center of our worldviews—as many of us already seem to be doing—we're lowering the bar on what we count as enough for our response to the climate crisis. We're deciding that as Earth warms, seas rise, biodiversity collapses, people move by the millions, and societies start to fall apart, it will be enough, simply, to give up those things that don't seem immediately important, rather than aggressively tackle climate change's underlying causes—especially our prodigious consumption of fossil fuels.

## BEDROCK VALUES

When we acquiesce to progressively greater abandonment of our values and others in the way just described, we seem to be sliding into a kind of moral abyss. How can we stop the slide?

Here's a good first step: identify a bedrock set of pertinent values and goals that most people around the world share, that we generally believe we won't compromise under any circumstances, and that can constitute the foundation of our vision of the future and the object of our hope.

Finding shared values might seem a tall order, given humanity's often quarrelsome behavior. But as noted in chapter 9, all people have the same basic physical needs for such things as healthy food, clean water, and freedom from disease, as well as the same basic psychological needs for safety, love, belonging, and esteem. And all parents want their children to be secure in the present and to have good prospects in the future. These needs are humanity's deepest shared values.

Yet it's precisely our ability as a species to continue satisfying these basic needs, after centuries of broad progress, that's now in doubt because of the colliding changes in our world—from the coronavirus pandemic to massive environmental damage to erosion of democratic norms and institutions. Of course, conditions are far more dire for some than others, just as in our metaphorical leaky lifeboat some people are closer to the leaks than others. Hunger, thirst, sickness, and war still stalk perhaps a billion of our fellow human beings today, and the numbers are starting to rise once more in both absolute and percentage terms. Such problems remain distant threats for others, however. Indeed, too many people with wealth seem to think they can wall themselves off from these threats—by retreating into the equivalent of gated communities in many areas of life. But if the dangerous changes we've unleashed continue unabated, all of us, the wealthiest included, will eventually find our basic needs compromised, just as everyone in the leaky lifeboat eventually drowns. A strategy of triage will merely postpone the inevitable.

When we look at humanity's profusion of big problems through a lens of basic needs and deepest shared values, they seem far less disparate and contentious. They start to look like an interconnected set of challenges to the well-being of everyone on Earth. Instead of being one big muddle that divides us, they start to look like common problems that might just bring us together around clear superordinate goals.

Again, I find a metaphor brings the matter into vivid relief. But in this case, the metaphor's key element isn't a boat but a house. The house is shared by a bunch of unruly young people who've had a lot of offspring before they're mature enough to look out for their children's interests properly. Together they thoroughly trash the place—ransack its cupboards for food, tear apart its walls for fuel to burn to keep warm, dump garbage from wall to wall—and then, as the necessities of life begin to run out, they angrily grab guns to kill each other if need be.[13]

It's a crude, disturbing image. But it seems to me to capture the essence of our situation today—our crazy behavior, considering the inescapable reality of our common fate on a shared planet. And it suggests three bottom-line injunctions for the future:

- Don't wreck our planetary home.
- Don't commit mass suicide by fighting among ourselves.
- Protect our children.

The first two are negative injunctions: fervent calls to refrain from doing certain things. The third is a positive injunction and to a certain extent is implied by the first two (and vice versa). Each of the three reflects, as I'll explain later, our fundamental moral duties.

I think of them as the floor—the absolute bottom line—of our aspirations for the future. We need to do *at least* this much to survive.

Arguably, of course, given the amount of global warming that's in train, we're already breaking through that floor when it comes to the first and third injunctions. But that reality doesn't mean we should give up on them. To my mind, these injunctions mark the limit of how far we should lower the bar on what counts as "enough." They're a tough reminder that responses to problems like climate change, or the danger of nuclear war, that won't produce outcomes satisfying all three conditions simply aren't enough.

Yet such minimum aspirations can't work as superordinate goals. They're not specific enough. Also, depending on how one defines "our" or "we," the third injunction doesn't require cooperation among the world's peoples, because while any one group might assent to protecting its own children, it might not assent to protecting those of others. And the injunctions' very defensiveness weakens their emotional power. It's hard to imagine how they'd excite any kind of broad collaboration across humanity to make things better.

Instead, they're a starting point for developing a larger moral framework and vision of a positive future that can become a motivating object of humanity's hope. While they leave many vital issues unaddressed—for instance, the place in our vision for modern commitments to human rights, equity, and democracy—their fundamental simplicity is a virtue, if we want to forge a broad consensus across humanity's diverse societies, cultures, and perspectives.

Contrast this simplicity with the tangle that's the "Sustainable Development Goals," announced by the United Nations in 2015.[14] Extruded like a string of sausages from an intensely bureaucratic and highly political process of global consultation, the seventeen SDGs, as they're called, are the best current statement of humanity's consensus on where we should be going as a species. They include items like "No Poverty" (number 1), "Quality Education" (number 4), and "Responsible Consumption and Production" (number 12). They're linked in turn to 169 policy targets, many of which specify measurable changes in people's circumstances that, it's hoped, should be achieved globally by 2030.

The ambition is noble, much of the detailed content in the UN's background documents for the SDGs makes eminent sense, and the SDGs firmly avoid any hint of triage. But the goals sometimes contradict each other. Most obviously, their commitment to continuous and rapid global economic growth (in number 8) is incompatible with their commitments to protecting the global environment (in numbers 6, 13, 14, and 15).

Perhaps their biggest shortcoming, though, is that they're simultaneously too elaborately technocratic and too blandly anodyne to truly motivate us. If we're going to pull together, we need principles and goals that combine to form an inspiring and coherent vision of the future that people everywhere can take to heart, affirm in conversation with each other and rally around, and then work with excitement together to reach.

I'm going to propose later several principles for humanity—and the goal of rebuilding nature—that could form the core of such a new, motivating vision. If enough of us around the world adopt them, we'll fundamentally change the *W* in the WIT—the set of worldviews, institutions, and technologies—that dominates humanity's global socio-ecological system. Aiming for such an outcome is extraordinarily audacious. But raw ambition must characterize any project with a real chance of being enough to put humanity on a truly better path.

# PART THREE

# THE PATH TO HOPE

# 14

# From Gondor to Washington, DC

*If you wanted to divert a mighty river into a different course, and all you had was a single pebble, you could do it, as long as you put the pebble in the right place to send the first trickle of water* that *way instead of* this. Philip Pullman

EVEN THOUGH HUMANITY HAS SEEN instances of major change in entrenched sets of worldviews, institutions, and technologies (WITs) in the past, today such change, which must include the development of an inspiring vision for humanity on a world scale, doesn't look feasible on first assessment, because so many forces are set to resist it.

By far the hardest transition will involve getting from today's economic growth WIT to another arrangement that drastically reduces the global economy's consumption of resources and its output of waste, while dramatically boosting the living standards of the world's poorest half. This new arrangement must explicitly address the three equivalencies I highlighted—growth equals happiness, freedom, and peace—because people won't relinquish conventional growth if they aren't reasonably sure they'll be at least as happy, free, and secure as they are under the existing arrangement.

The intellectual and scientific foundation of this new WIT will also need to incorporate a renovated discipline of economics—one that recognizes that human economies are complex systems intimately connected with nature; that markets won't automatically find good substitutes for some of the most precious things nature gives us, like moderate temperatures and enough water for our crops; and that economics must be grounded in moral principles attuned to our world's demanding new material and social realities. And, to top it all off, our alternative economic WIT should avoid the suffocating burden of increasingly complex government regulation, while retaining the technological creativity of modern capitalism within a democratic framework.[1]

Okay, sure. Let's just do that.

It's difficult to imagine even the macro features of an alternative arrangement that meets all these conditions—although some researchers have made progress describing these features—let alone the details of that alternative or a feasible path to get us to it.[2] Still, I'm sure about two things: any alternative economic WIT that's enough will be radically different from what we have around us now; and as today's economic growth WIT starts to break down, several other partially formed WITs will compete to replace it. The process of breakdown has clearly begun, and it's already proving violent, in the form of increased racism within many of our societies as economic insecurity and inequality rise; vicious attacks on minorities and, in more and more countries, on environmental activists; and huge internment camps metastasizing along national borders to hold migrants and climate refugees desperately looking for better lives. It's also catalyzing broad assaults on basic principles and institutions of democracy, not just within relatively new democracies such as Hungary, Poland, Brazil, and India, but even in one of the world's oldest democracies, and certainly its most important, the United States.

I'm not at heart a revolutionary. I'm also enough of a historian to know that few processes of violent social change leave things better than they were before. To reach a positive future, we need to find ways to shift today's dominant WITs with as little trauma and violence as possible. Rather than trying to overthrow these WITs with head-on, brute-force efforts, we should look for their weak spots to achieve the radical change needed.

### VIRTUOUS CASCADES OF TRANSFORMATIVE CHANGE

The trick is to exploit the enormous leverage that's available in our highly nonlinear, global socio-ecological system. We know that small changes in nonlinear systems can sometimes cascade into huge system-wide shifts. So perhaps, if we can find and exploit the right intervention points in the web of links between our worldviews, institutions, and technologies—and if we can take advantage of our world's connectivity and uniformity, both factors that can multiply interventions' effects—we might be able to use targeted actions to trigger what I like to call a *virtuous cascade* of change that flips humanity onto a more positive path, without massive violence or trauma.

That sounds like a tall order, but a short, personal essay written by the renowned environmentalist and systems theorist Donella Meadows, not long before she died in early 2001, suggests where we can start. Meadows led the game-changing 1972 *Limits to Growth* study, and this single essay is her distillation of a lifetime's expertise on how systems work. Titled "Leverage Points: Places to Intervene in a System," it has become a reference point for many analysts concerned about humanity's fate and looking for a path forward. And although almost two decades old, it reads as if written yesterday.

Donella Meadows opens with these words:

> Folks who do systems analysis have a great belief in "leverage points." These are places within a complex system (a corporation, an economy, a living body, a city, an ecosystem) where a small shift in one thing can produce big changes in everything.
>
> This idea is not unique to systems analysis—it's embedded in legend. The silver bullet, the trimtab, the miracle cure, the secret passage, the magic password, the single hero or villain who turns the tide of history. The nearly effortless way to cut through or leap over huge obstacles. We not only want to believe that there are leverage points, we want to know where they are and how to get our hands on them. Leverage points are points of power.[3]

Meadows identifies a dozen such points where people could do things to flip a socio-ecological system from one state to another. She lists them from top to bottom in order of increasing "effectiveness," with the more effective leverage points (towards the bottom of the list) able to generate larger system shifts (see the adjacent table).

Two of her conclusions stand out. First, less effective leverage points (those towards the top of the list) generally change things like a system's flows of matter, energy, and capital. So, they're located mainly in the system's material components—in those parts that exist mainly in the physical world. Her more effective leverage points, on the other hand, are located mainly in people's heads and concern things like information flows and our rules, goals, beliefs, and values—what social scientists call "ideational" components. (Her most effective leverage point—"the power to transcend paradigms"—might seem obscure, but I'll show it's her most brilliant insight.)

Second, if we imagine an actual lever at each of Donella Meadows's points—in the Archimedean sense of a lever that pivots

## PLACES TO INTERVENE IN A SYSTEM
### IN ORDER OF INCREASING EFFECTIVENESS

*(according to Donella Meadows)*

12. Constants, parameters, numbers (such as subsidies, taxes, standards)
11. The sizes of buffers and other stabilizing stocks, relative to their flows
10. The structure of material stocks and flows (such as transport networks, population age structures)
9. The lengths of delays, relative to the rate of system change
8. The strength of negative feedback loops, relative to the impacts they are trying to correct against
7. The gain around driving positive feedback loops
6. The structure of information flows (who does and does not have access to what kinds of information)
5. The rules of the system (such as incentives, punishments, constraints)
4. The power to add, change, evolve, or self-organize system structure
3. The goals of the system
2. The mindset or paradigm (worldview) out of which the system—its goals, structure, rules, delays, parameters—arises
1. The power to transcend paradigms

on a fulcrum—her least effective levers seem to be easiest to access and "pull," whereas those she designated as most effective seem to be hardest to access and pull. In other words, she suggests that there's an inverse relationship between a lever's potential effect and the degree to which we can use or exploit that effect. In the terms we've used here, more effective levers, if pulled, are more likely to get us to an outcome that's enough, but pulling those levers is less feasible.

Donella Meadows was an expert in system dynamics, the kind of mathematics used in the original *Limits to Growth* study to model the flow of materials and energy in natural and social systems. While in her article she didn't explicitly use the notion of social power that I introduced in chapter 10, it's clear she implicitly understood this power's role in blocking positive change, especially when used by vested interests. In discussing her fifth most-effective leverage point—the "rules of the system" (equivalent to what we've called "institutions")—she writes:

> Power over the rules is real power. That's why lobbyists congregate when Congress writes laws, and why the Supreme Court, which interprets and delineates the Constitution—the rules for writing rules—has even more power than Congress. If you want to understand the deepest malfunctions of systems, pay attention to the rules, and to who has power over them.[4]

### "HOUSEWIVES" USING LEVERS

The Donella Meadows leverage-point framework is a wonderfully useful tool to see deep into the inner workings of complex systems and to pinpoint how to change their behavior. It even sheds new light on fables like *The Lord of the Rings*.

A key part of that story involves the reinvigoration of a moribund Alliance of Men. Generations earlier, the kingdoms of Rohan and Gondor had concluded a defensive pact that could be invoked if either side were to light a series of beacon fires strung across the vast landscape between the two lands. But distance, conflicting interests, and the evil Lord Sauron's machinations have sown distrust between these kingdoms and their leaders; and, as a climactic battle against Sauron's forces approaches, it's not at all clear they'll come to each other's aid when called. So the wizard Gandalf reminds the king of Rohan of the alliance's moral imperative and, later in Gondor, orders the Hobbit Pippin to climb the nearest beacon tower, at grave risk to his life, to light the fuel at its pinnacle—even though the leader of Gondor has fallen under Sauron's sway and doesn't want Rohan's help. The signal is sent, Rohan dispatches its army to Gondor, and the tide of battle is ultimately turned against Sauron.

In complex-systems terminology, the alliance between Rohan and Gondor is a latent "negative" (or stabilizing) feedback loop: should the social system in which Men live—the societies of Rohan and Gondor taken together—be threatened or destabilized, the alliance will help return it to a secure equilibrium. But because it's not clear that this feedback is still capable of operating, Gandalf bolsters it in his conversation with Rohan's king; he then removes an impediment to it by circumventing Gondor's leader. Gandalf, in effect, manipulates "the strength of negative feedback loops"—Meadows's eighth most-effective leverage point. And when he lights the first beacon, Pippin provides a marvelous example of how exploiting the right intervention point can precipitate an enormous cascade of (in this case, good) consequences.

Well, that's just a fable. What about a real-world example of people exploiting leverage points? Stephanie May, it turns out, had a remarkable intuition for how to multiply the impact of her campaign against nuclear testing. Rebounding quickly from the disappointing reaction to her idea for a Lobby for Humanity, through the first half of 1957 she wrote a stream of letters about nuclear testing to local and national newspapers. She had a knack for making her points—through brevity, force of moral argument, and often wit—in ways that editors were eager to publish.

In January 1957, for example, the Associated Press reported on testimony to the US Senate by Warren Weaver, the chairman of the Committee on Genetic Effects of Atomic Radiation of the National Academy of Sciences. He estimated that radioactive fallout would cause an additional six thousand babies—born to the current generation of mothers worldwide—to come into the world with mental and physical handicaps. This was, he averred, "a very small fraction of the total" of babies; and later to reporters, he said he thought the harm was a "fair price" for the military security provided by nuclear testing.[5] Less than a week later, under the heading "Fair Exchange?" the *New York Post* published Stephanie's response in its letters to the editor section:

> Will the hearing on Genetic Effects of Atomic Radiation call in the mothers of these 6,000 handicapped children to ask if they think the price they have paid for nuclear tests is "fair"?[6]

She knew that for outlets of national importance like the *Post*—in those days, a renowned liberal tabloid with a lineage going back to its founding by Alexander Hamilton in 1801—she had to make her point fast and powerfully. On June 1 she succeeded again when, under the heading "Gambling with Humanity," the *Post* published the following letter from her:

> Does the AEC [Atomic Energy Commission] hold its tests in Nevada because gambling with humanity is legalized there?[7]

About the same time, she received a handwritten note from a housewife in Storrs, Connecticut. Virginia Davis had recently arrived from California, where she had been doing regular news commentaries on her town's radio station on the dangers of testing. She'd seen Stephanie's letters in a local paper and suggested the two of them meet to "form a committee" in their area to oppose nuclear testing. Barely four weeks later, with two thousand signatures in hand, Stephanie and Virginia took a bus to Washington, DC, to deliver the petition to the White House—a delivery facilitated by their local Republican congressman, Edwin May (no relation to Stephanie), who just happened to golf with President Eisenhower.

Before they delivered the petition, they roamed the halls of Congress with the document in hand—something inconceivable in our security-conscious era—using it to open doors and begin conversations with some of the country's most powerful politicians. By the end of their first day, they'd met five congressmen and six senators, including senators Jacob Javits, Albert Gore (the father of the later vice president and presidential candidate), and Connecticut senator Prescott Bush (father and grandfather of the later presidents).

In her memoirs, Stephanie recounts a charming anecdote about how she and Virginia asked a young Senate page to call Hubert Humphrey from the Senate floor. When the senator emerged, he noted "it's really dull in there" and thanked them for giving him an excuse to stretch his legs.[8] After they'd talked about nuclear testing and the petition, the senator invited the women to discuss the matter later with his disarmament assistant. The next day, the *Washington Post*—oblivious, as were almost all mainstream media outlets in those days, to its offensive, patronizing tone—featured a story titled "Housewives Petition Ike to Halt Nuclear Testing," with a photograph of Stephanie and Virginia holding the petition's sheaf of papers.[9]

Many setbacks followed in the years to come, but these two days were a significant moment in the worldwide campaign against atmospheric nuclear testing.

In Donella Meadows's scheme, Stephanie (with her letters to the editor), and later Stephanie and Virginia (with their petition and trip to Washington, DC), exploited and manipulated "the structure of information flows" in the social systems around them (leverage point number six). This is a system lever that political activists of all stripes—past, present, and future—always use. "Missing [information] feedback is one of the most common causes of system malfunction," Meadows writes. "Adding or restoring information can be a powerful intervention, usually much easier and cheaper than rebuilding physical infrastructure."[10] The two women single-handedly created an information link from worried mothers to the country's political elites. In the years to come, Stephanie made sure that the link constantly pulsed with information, so that no powerful politician—conservative or liberal, Republican or Democrat—could remain unaware of the mothers' passionate opposition to nuclear testing.

### THE CRUCIAL PLACE TO START

Where should we cross the boundary into the vast adjacent possible? In what direction are we most likely to discover or create a feasible path towards a future that's enough?

Taking a hint from Tolkien, perhaps we should look for solutions in the opposite direction of what seems initially to make the most sense. This would mean downplaying, even ignoring, the conventional wisdom of our policy elites, which says that when we're trying to address problems like climate change, rising economic inequality, or the danger of nuclear war, we should pursue what's politically, economically, and socially feasible over what would be enough. And that business-as-usual approach almost always means our route should lie through incremental tweaks to existing institutions or through new technologies. Yet, to echo Tolkien's ominous language, it's becoming abundantly clear that this route, if followed exclusively, will lead us to our doom. As the young climate activist Greta Thunberg told the assembled dignitaries and scientists at the United Nations' climate conference in Katowice, Poland, on December 12, 2018: "Until you start focusing on what needs to be done rather than what is politically possible, there is no hope."[11]

Donella Meadows's leverage-point model suggests an alternative way forward.

The items in her list fit easily within the worldview-institution-technology (WIT) framework. Some of her places to intervene concern mainly technologies, while others concern institutions; but the three she believed were most effective all relate to worldviews. (By "paradigms," Meadows meant worldviews: she called paradigms our "mindsets" or our "deepest set of beliefs about how the world works.") Since a more effective leverage point is more likely to get us to an outcome that's enough, her model implies that if we want to reach a desirable future, we'd be wise to give much greater priority to possibilities of worldview change.[12]

Yet she also suggests that more effective leverage points are generally harder to exploit—that, in our words, taking advantage of such leverage points is usually less feasible. But I'm not sure she's right when it comes to worldviews. For one thing, each of us can explore alternative worldviews immediately in our individual minds, to see how other perspectives look and function from the inside and whether aspects of them seem better than our current perspectives.

Even Meadows acknowledges that "[there's] nothing physical or expensive or even slow about paradigm change. In a single individual it can happen in a millisecond. All it takes is a click in the mind, a new way of seeing."[13] So individual worldview change is entirely feasible.

When so many other things in our world—most obviously the massive institutions and technological systems we're embedded in—seem impossible for any single one of us to shift, even a millimeter, it's heartening to know that we can still exercise our agency directly and palpably over our individual worldviews. We can, each one of us, take our own first steps into the adjacent possible, at the point along the boundary where worldview change starts.

Yet these changes mustn't stay just in our individual minds. We must share them with each other. The mental tools I'll introduce in the next chapters are designed to help us explore the adjacent possible of alternative worldviews together and talk more effectively with each other about what we discover there. We've already seen how human beings can create new aspects of their social reality simply by believing that they're real, communicating those beliefs among themselves, and then acting concretely together on them. So our *social* future is radically open if we imagine, create, and act together—true novelty is indeed possible, although our social inventions, to survive and thrive, must effectively recognize our world's increasingly constrained material reality, including the slow-process constraints we identified earlier—demographic imbalances, climate change, declining biodiversity, falling resource quality, and the like.

Of course, the point is ultimately to do something about—not just to think together in new ways about—those material realities that are pushing us to calamity. And it's true that because our worldviews, institutions, and technologies are tightly connected in self-reinforcing systems, we must act on all three fronts more or less simultaneously if we want to produce real, positive change in those realities.

I believe that a better understanding of our own and others' worldviews is the crucial place to start, however. It can help us create new social and political alliances among groups around the world; then, within these alliances, we can work together to build a scaffolding of concepts, beliefs, and values that supports the necessary transformation of our planet's institutions and technologies, particularly those that govern its economic system. The better understanding will also empower us to circumvent or even rout groups opposing this transformation.

Exploiting the worldview leverage point is more feasible today than Meadows believed twenty years ago, because our modern world's extraordinary connectivity and uniformity can amplify the speed and extent of a global worldview shift, once it starts to happen.

It's worth recalling that technological change generally happens more slowly than we expect, while social change can happen astonishingly fast. As noted earlier, the most visible aspect of such fast change is usually a major institutional shift—the collapse of the apartheid regime or the sudden demise of Soviet communism, for instance, or more recently the legalization of gay marriage in many Western countries. But those shifts occurred only because substantial worldview change had already taken place under the surface, often—as in the cases of apartheid and gay marriage—due to the efforts and courage of many, many people at all levels of society.[14] A large proportion of the population had already stopped believing in the dominant worldview that supported the existing institutions.[15] People continued, for a while, to *act* in their communities as if they believed the dominant worldview, because acting otherwise might have been harmful in some way. But beneath the surface many people's attitudes had changed and that change ultimately enabled a seemingly abrupt institutional shift.

Today and in the future, the underlying process of worldview change could be far more visible and greatly accelerated. Intentional and targeted shifts in our worldviews could propagate through our planet's densely linked social and information networks and empower changes in politics, institutions, and technologies—in swift, nonlinear jumps towards our goals. So in the next chapters I'm going to focus on how we can improve our understanding of worldviews—how they're structured, how they function, and how they evolve. Because I'm convinced that this is our best place to cross the boundary into the adjacent possible, and to find a feasible route to a future that's enough.

# 15

# Into the Mind

*Nothing is more frightful than to see ignorance in action.* Goethe

IN PLACE OF THE FEELING of boundless possibility that animated many societies towards the end of the twentieth century, many people around the world now increasingly see their horizons as far nearer and their range of future opportunities as far narrower than before. In this critical moment of our species' evolution, our worldviews should instill in us all feelings of promise and possibility that inspire us to pull together to address our unprecedented problems. But much of humanity seems to be heading in the opposite direction—towards truly baleful or depressive attitudes that will just make our problems worse.

Global polling firms have tracked this change. Their data tell an inescapable story. And it's worth revisiting some of the basics of this story before examining how we got to our current psychological state—bitter about our prospects and often pitted against each other—so we can gain some insight into how we might best respond.

In early 2019, the Pew Research Center reported that pessimism about children's economic future had become "widespread in most economies," with 56 percent of people polled in advanced economies, and 53 percent in emerging economies, saying that today's children in their countries would be "worse off" financially when they grow up than their parents.[1] The Ipsos polling firm, surveying attitudes in 2016 across twenty-three countries—including Argentina, Brazil, Germany, India, South Africa, and the United States—found that 48 percent of those polled believed today's youth would have a "worse life" than their parents' generation, with only 26 percent feeling youths' lives would be better.[2] Ipsos also found that just 28 percent of people overall, when they looked ahead a year, were optimistic about prospects for the "world in general."[3]

There are understandably large differences in attitudes between countries though. Generally, people in emerging economies with robust growth—like China, India, and Indonesia—feel far better about the future for themselves, their children, their country, and the world. For instance, while only 18 percent of those polled in established economies were optimistic about the world in general, 42 percent were optimistic in emerging economies. But in the West especially, the shift in social mood since the beginning of the millennium has been stark. According to an Ipsos MORI study conducted for the British newspaper the *Guardian,* between 2003 and 2016 the optimism-pessimism balance in Great Britain almost completely flipped. In 2003, 43 percent of people polled thought that youth at the time would have a "higher/better" quality of life than their parents, while 12 percent thought youth quality of life would be "lower/worse." But by 2016—even before the Brexit fiasco had caused political and economic turmoil in the country—only 22 percent thought youth quality of life would be higher than their parents, while 54 percent thought it would be lower.[4]

What has caused such a remarkable shift? Many factors are operating, and some are unique to each country in question. But I'm sure people are reacting, at least in part, to the early hints of the enormous social earthquakes our societies will likely undergo in coming decades, as hard-to-see, slow-moving, and diffuse tectonic stresses steadily build in force, cross social boundaries and scales, and combine to multiply their effects. Four of those stresses (three of which I mentioned in chapter 9) seem to be having an outsized impact on people's moods, especially in the West.

The first is a combination of widening economic inequality and rising economic insecurity—that is, a fear that one's basic economic well-being could change abruptly for the worse at any time, for instance from loss of work. Both inequality and insecurity were made worse by the more sluggish pace of global economic growth after the economic flip in 2008–09, and were then dramatically exacerbated by 2020's pandemic-catalyzed economic downturn. In many countries hard hit by the 2008–09 financial crisis, average incomes had barely returned by 2020 to where they were previously (and in some communities they never fully recovered). Meanwhile, overall economic and social inequality continued to rise relentlessly at the same time that job security declined.

The people hurt most by these trends are those who earn their living mainly through physical work in fields and factories—often outside major urban zones—or in retail service industries. Since the last century, far greater economic benefit has flowed to those who generate or manipulate ideas and information in office buildings in downtown cores. People on the losing side of this divide feel more and more that the economic system is inescapably unfair to them and that well-connected, well-healed elites can bend this system's rules in their favor.

The second stress affecting our collective mood is the growing movement of people, chiefly economic migrants and refugees, from areas of the world where life is terribly hard and dangerous to areas where it could be better. People rarely leave their homelands or countries en masse unless something is very wrong. Today's migrants are usually fleeing rural regions where livelihoods are undermined as soils, fresh water, and forests are degraded and weather becomes more extreme; cities whose economies are in crisis and where jobs are scarce; and zones, both rural and urban, where gang, terrorist, ethnic, or government-sponsored violence is rampant. Societies on the receiving end of the streams of migrants may welcome the newcomers generously at first (as Germany did in 2015), but over time, if the number of people arriving doesn't fall sharply, the welcome mat is almost always withdrawn, usually when opportunistic politicians exploit people's fears of outsiders.

The third stress is climate change, which is something of a stealth threat to people's feelings of security, possibility, and hope. To the extent we're aware of the problem, we might try to help by making some changes in our everyday lives—recycling more or riding a bike to the store instead of taking a car—but as I've noted, the sheer size and complexity of the issue leaves many people sensing that any personal lifestyle changes are futile. Most of us also see that with a few notable exceptions, the world's political leaders hide behind excuses. Only occasionally, as in the United States under the Obama administration, have our leaders proposed policies that really challenged the economic status quo in this regard.[5] Meanwhile, evidence that climate change is getting steadily worse—including evidence that people can now feel and see, like wildfire smoke—continues to pile up.

The fourth and final stress that's having a major effect on humanity's collective mood, one that I haven't mentioned yet, is called by some scholars "normative threat."[6] It arises partly from economic inequality and insecurity and perhaps from the arrival in one's homeland of large numbers of migrants, but also from rapid urbanization and greater information connectivity, both of which allow ideas to propagate quickly, especially new beliefs and values about traditionally contentious issues like sexuality, religion, and women's status. Together, these changes can make some people feel as if their society's essential fabric of culture, moral values, and shared beliefs, myths, and practices—the fundamental "normative order" that constitutes their society's continuity—is being torn apart. Polls register this feeling, too: around the world, 79 percent of people agree that "the world today is changing too fast," according to Ipsos; 58 percent think that "society is broken"; and 57 percent believe their own country is "in decline."[7] And once again, opportunistic politicians, Donald Trump and the United Kingdom's Boris Johnson being prime examples, can capitalize on these sentiments to gain attention and power.[8]

These four slow-process stresses—worsening economic inequality and insecurity, mass migrations, climate change, and normative threat—erode our feeling that our situations are safe and fair; such stresses are also seen by many people, even if only subconsciously, as harbingers of devastating social earthquakes to come. An astonishing 82 percent of people surveyed worldwide by Ipsos think we live in an "increasingly dangerous" world.[9] Yet, our conventional leadership, and bureaucratic and corporate elites, seem wholly incompetent, impotent, or indifferent in response; and our institutions seem largely unable to cope.

And it's these perceived failures that are at the root of humanity's deep shift in social mood. They encourage people to flip from what the *New York Times* columnist and public policy author David Brooks calls an "abundance mindset" to a "scarcity mindset." In a scarcity mindset, "Resources are limited. The world is dangerous. Group conflict is inevitable. It's us versus them. If they win, we're ruined, therefore, let's stick with our tribe. The ends justify the means."[10]

Techno-optimists find this mood shift utterly perplexing. Why, they ask, are so many people today so glum, when circumstances are patently so much better for most people than they were, say, fifty years ago? The answer is very simple, really: we know that people's social mood is far less affected by where they've been in the past than by where they think they're going in the future. A sour view is particularly likely when the future doesn't appear nearly as good as the not-distant past. And that's what many folks see now when they look forward, especially in the West: a lot of trends heading in a decidedly wrong direction. Yes, some cynical politicians exaggerate the negativity of these trends to further their narrow ends, but these exaggerations usually exploit underlying and dispiriting realities.

Ironically for techno-optimists, their vaunted information revolution has played a key role in propelling this shift. Barely fifteen years ago, it was widely believed that societies wired together by the internet and the web would become progressively smarter over time—that a higher collective intelligence could emerge from rapid flows of immense amounts of information and a flattening of knowledge hierarchies as everyone gained direct access to previously inaccessible expertise.[11] Since then, we've learned some harsh lessons. Instead of creating a digital environment that draws us together and makes us smarter, the companies at the core of the social media revolution—Google, Facebook, YouTube, Twitter, and the like—have used the vast amounts of data they harvest about our preferences and behaviors to create an emotional environment that tends to pull us apart and make us dumber. These companies get more eyeballs on their sites and clicks on their links—and so generate more revenue—by creating an addictive information environment of constant cognitive and emotional stimulus, and among the emotions that work best are outrage and (social) anxiety. Such emotions make us feel chronically insecure, encourage us to coalesce into groups of the like-minded for psychological protection, and dissolve the trust between diverse peoples that's essential for a consensus on what's true and untrue. They also subtly erode our faith in the future.[12]

Already for most people in the West, the ideal of social and technological "progress," which just twenty years ago we widely accepted without much reflection, seems oddly anachronistic; while the once-bright ideal of Western-style capitalist democracy guided by reason and science, which the scholar Francis Fukuyama boldly declared was the blissful endpoint of human political evolution, is fading fast. These changes signal that many of us are losing confidence in a positive future.[13] If we lose it entirely, our aspirations will drastically narrow; some of us will become embittered, detached from others, resigned, and morally immobilized; while others may flee into political or religious zealotry, grab-what-you-can self-interest, or, at the other extreme, live-for-today partying. And if most people lose hope in a positive future, humanity will slide into ever more bitter quarrels among groups holding what I called earlier "Mad Max" worldviews—quarrels that will undermine any possibility of a shared commitment to the species' broader commonweal, and with it the possibility of genuinely reducing the dangers we collectively face—of achieving, in other words, something resembling "enough."

### STUPID OR DECEITFUL

Fear is doing much of the dirty work here. It's a useful emotion—as when it motivates us to escape from danger or fix the underlying problems that are its causes. After all, higher organisms evolved the biological capacity for fear because it helped them survive, which allowed them to pass their genes to future generations. Yet while a certain degree of fear can help us stay resilient in an ever-changing and often hazardous world, if fear is too great, and if we see no avenue for relief, it becomes horribly toxic. "No passion," said Edmund Burke, the English statesman and philosopher, "so effectually robs the mind of all its power of acting and reasoning as fear."[14] Most importantly, persistently high levels of fear tend to divide people into shortsighted, rigidly exclusive, and antagonistic groups.

But why are so many people nowadays so afraid? What is it about the rising stresses in our world that so scares us? Part of the answer, of course, is that countless people have strong material cause to be afraid—their employment is more precarious, the political power of their social groups is waning, their medical costs are rising (especially in the United States), their social support networks are fraying, and on and on. But another key part of the answer, I believe, lies in how the stresses threaten vital elements of our worldviews. Because our worldviews satisfy some of our most fundamental psychological needs, when they're seriously threatened, we can get very scared.

This psychological process is visible in the bruising, often-vitriolic debates around climate change. There's something about the issue that can make reasoned conversation formidably hard and even turn people into bitter opponents.

We've all had firsthand experience with this kind of dispute. We've become embroiled in a row with someone about a matter of politics, business, science, or ethics and then been frustrated, flabbergasted even, as we've discovered we can't convince the other person we're right. We're sure we've got unassailable facts and logic on our side, but these advantages just don't seem to matter. Every point we make runs headlong into a rebuttal or counter-argument. Some of our interlocutor's responses are simply wrong, we are sure, while others might be true but unimportant, and still others seem to be deliberately diversionary. We do our best to answer each in turn, yet after a while we conclude the whole conversation is pointless.

The other person just doesn't "get it." Our seemingly excellent reasoning has run into an impervious wall of rejection.

Such arguments can even turn family gatherings into truly miserable events. Most of us learn to avoid mentioning certain subjects in the interest of harmony—politics, or values around gender, or racial and immigration matters. But sometimes we need at least a partial meeting of minds to find a workable solution to an urgent or dangerous problem, or when the problem is so complex that no one has a monopoly on wisdom about how to respond.

This is exactly the situation with climate change. For three decades, I've written and spoken about the issue to a broad range of audiences. Over those years, as the evidence of warming has become more palpable in the form of events like droughts, storms, and wildfires, the public in most societies has increasingly accepted the basic science of climate change. In late 2018, for instance, 73 percent of Americans agreed global warming is occurring, and 62 percent said humans were the main cause.[15] Yet at my talks, people in the audience remain divided into three basic categories: some are receptive to the arguments, others pay attention but remain wary, and yet others are downright hostile, rejecting almost entirely what I offer about the findings of climate science and how I believe humanity should respond. Sometimes when I speak to an audience of folks in the fossil-fuel industry, the moment I mention the scientific consensus on climate change, nearly everyone in front of me crosses his or her arms.

Perhaps I'm a glutton for punishment, but I often engage with people who profoundly disagree with me on global warming—let's call them "climate change contrarians." (There's a debate about the appropriate label for people who don't accept the broad scientific consensus on climate change, its likely consequences, or how it should be addressed. The most common terms are "skeptics," "deniers," and "contrarians." Each has strengths and weaknesses. I prefer the last, because these folks' views are contrary to the scientific consensus.) I do so for three main reasons. Partly, of course, I want to see if I can change their minds. While the balance of public opinion has decisively shifted against contrarians in recent years, they remain enormously powerful in many societies, especially within elite conservative circles, where they're often key actors in "blocking coalitions" of vested interests stopping, or even rolling back, action on carbon emissions. So I want to see what arguments and evidence might persuade some contrarians to change sides. Secondly, as an academic and social scientist, I'm committed to the vital role of reason in human affairs, and I regard my conversations with contrarians as little case studies into why people (including me!) sometimes have so much trouble rationally working out their deepest disagreements. And finally, but not least, I know too that I might learn something from talking to people who strongly disagree with me; maybe my views are wrong in some respects.

Yet it's not easy, as we all know, to really see and engage with the "other side"—whatever that side may be. A concession in the argument can lead to more open discussion, but just as often it's like blood in the water to a shark. I've often come away from these conversations deeply frustrated!

I admit that, when I was younger, I routinely concluded after my encounters with climate contrarians, as one might after a heated family argument, that the folks on the other side were either a bit stupid or a bit deceitful—or maybe even a lot of both. Either they couldn't see the scientific facts about climate change in front of their noses, or they were deliberately and consciously dismissing these facts for self-serving reasons, perhaps because of, say, a job in the fossil-fuel industry. In either case, as an ideological liberal, I was sure back then that right-wing ideology played a vital role and that my contrarian interlocutors were often hiding their blindness behind shibboleths about nature's resilience or the power of markets: "Nature is so vast that we can never really hurt it," or "capitalism can solve this problem when it gets bad." Other times they seemed to be using capitalist ideology to convince themselves of the moral legitimacy of their position: "If we're to bring people forward economically in all societies, protecting economic growth is more urgently important than doing something about climate change, which in any case probably isn't even happening," I heard (or imagined) them saying.

Of course, a "stupid or deceitful" verdict about those who don't agree with us is easy to reach and satisfying to hold, because it absolves us of responsibility for the dispute. And once we identify others as the cause of the disagreement, we can blame them for why nothing is being done about the problem in question.

Yet on reflection, I had to grudgingly admit that many of my contrarian interlocutors were smart and well-informed, often citing scientific studies in detail and making carefully calibrated arguments. And it seemed hard to justify the judgment that those who took the time to talk to me at length were being deliberately deceitful.[16] I also realized that my interlocutors were interpreting my arguments much as I was interpreting theirs: in their eyes, I was either extraordinarily dense or deliberately twisting the facts about climate change; left-wing ideology had either addled my mind, or I was conveniently using climate change to justify a self-serving attack on the capitalist system.

Our views were in some respects mirror images of each other.

### PASCAL'S WAGER

Recently, a whole cottage industry of academic analysis has tried to explain what's going on here.[17] This research, along with my own experiences and study, suggests that the causes of the divide over climate change—and of similarly bitter divides over many other moral, political, and policy issues—are seldom disagreements over facts or even moral values, but instead are buried far down in people's worldviews and specifically in their deepest motivations for holding their worldviews. If true, the implications are vast.

Two thought experiments can help us think through these issues.

With the first, we can explore the worldviews of climate contrarians. We start by assuming we're reasonably well-informed about current climate science, as are many who dismiss climate science. Then we ask ourselves: What would we have to additionally believe to find it reasonable to argue that little or nothing should be done to address climate change's underlying causes?

Those of us who are worried about climate change are astounded that contrarians can take such a position. It's incomprehensible that to avoid the costs of reducing carbon emissions in the present, they seem prepared to risk the future well-being of today's children—even that of their own children. Do they not care about what happens in the future? Yet most genuinely say and believe they care a lot, and they're not lying. Indeed, some say that doing little or nothing to cut emissions will protect our children's future best, because the money thus saved can be invested in creating stability and wealth that our children will enjoy later in their lives.

On the other hand, though, given that mainstream climate science, backed by a near-consensus of scientists and supported by evidence from literally hundreds of independent lines of research, tells us that climate change will likely be catastrophic for humanity if nothing is done; and given the tangible reality of science's astonishing value and predictive power in other parts of our everyday lives—like every time we use a car, rely on an airplane's navigation system, or take antibiotics—how, we ask, can contrarians think that doing little or nothing is best? Their decision seems preposterously unwise to those of us who believe the science. And that's exactly why it's so interesting.

Academics who specialize in decision theory sometimes call the kind of choice we face with climate change a Pascal's Wager. Blaise Pascal, the brilliant seventeenth-century French mathematician and philosopher, famously described a wager on the existence of God. Under one interpretation of his argument, if we're faced with such a wager, we should bet that God indeed does exist, because betting that God doesn't exist—and then being wrong—will invite God's wrath and consignment to hell for eternity. That horrible consequence, no matter how improbable, should override any skepticism we might harbor about God's existence.

When it comes to climate change, humanity arguably faces a similar situation. On one hand, we can wager that the scientists are right about the potentially catastrophic results of climate change and so act aggressively to address the problem. If, when the future unfolds, we win the wager (that is, if the scientists turn out to be right that climate change would have been catastrophic if we'd left it unaddressed), the actions we've taken will have prevented catastrophe. If we lose the wager (that is, if the scientists turn about to be wrong about the danger), we'll have wasted some time and money addressing a non-problem.

On the other hand, we can wager that the scientists are wrong and so do nothing to stop climate change. In this case, if we win, we won't have wasted any time or money; but if we lose, we face climate catastrophe—the analogue of Pascal's eternity in hell.

Put this way, it seems blindingly obvious what we should do. As in the original version of the wager, the possibility of catastrophe must drive our decision. If there's any real chance—even a small one—that the scientists are correct about the potential danger, then we should place our bet on acting aggressively to prevent catastrophe—especially if, as advocates of such action almost universally believe, doing so won't badly damage our economies. It's like buying insurance to cover the small possibility our house will burn down, which is behavior we regard as part and parcel of normal prudential planning. And in the case of climate change, the argument for prudence is even stronger, because scientists say the risk of catastrophe if we do nothing is not small—it's extremely large.

Because this conclusion seems blindingly obvious to people who want action to stop climate change, they sometimes trot out the Pascal's Wager argument as if it were final and decisive. But when presented with such reasoning, most climate contrarians just shrug; they don't find it remotely persuasive, let alone decisive.

We can use our thought experiment to see why they remain unpersuaded, but it quickly takes us down a rabbit's hole of weirdness; everything becomes oddly distorted and strange. Scholars have long argued about how best to understand the underlying logic of the wager's original version about God.[18] The climate-change version is even more complex.[19] For instance, let's say we want to figure out how much we might pay if we wager climate scientists are wrong and so do nothing about the problem—and then it turns out the scientists were actually right. Humanity could then face consequences like monsoon failures in Asia, crippling droughts in the US Southwest, and raging wildfires in the Amazon. To estimate the total impact of these events, we'd need to estimate and then add up their individual impacts, which means we'd need to judge how much each would cost our societies if it happened, its likelihood of happening, and when it might happen.[20]

Dozens of technical assumptions underpin such estimates and calculations—about things like future energy technologies, rates of economic growth, the climate's behavior, ecosystem stability, social vulnerabilities, and people's values. The debate quickly bogs down in arcane details. But what becomes clear, after a while, is that these details don't really matter. For ardent contrarians, if the scientists aren't wrong about one thing, then they're surely wrong about a bunch of other things—easily enough, in their view, to discredit the claim that humanity faces catastrophe on its current course.

But that conviction can't stand by itself. By helping us step into contrarians' minds, the thought experiment shows it almost always rests on two deeper and more pernicious beliefs. First, because climate scientists clearly aren't stupid—they're trained scientists, after all—they must know what they're saying is wrong, which means that they're being deliberately dishonest. Second, because the climate scientists the world over largely speak with one voice about the dangers, they must be colluding in their dishonesty. (The most vocal contrarians claim that climate scientists are engaged in a broad left-wing conspiracy to terrify the public, raise their profile, and bilk governments, research agencies, and foundations of research

*Poster published in 2014 by the Heartland Institute, a staunchly contrarian US lobby group*

funding. I've heard all too often something like: "Well, you know, these guys need to exaggerate the danger to get their research grants renewed!") The assumption of collusion demands an extraordinary cognitive leap, but it also delivers one enormous advantage: if climate scientists are profoundly corrupt, then their scientific findings are thoroughly corrupted too, and one really doesn't have to debate the details. They can be dismissed wholesale.

Now those of us who aren't contrarians can see how we'd have to change our worldviews to find it reasonable to argue that, given the scientific information available, little or nothing should be done about climate change: we'd have to assume that almost all climate scientists are colluding liars. We can also now see why these folks find the Pascal's Wager argument unpersuasive. The decision to do nothing in response to the putative crisis of climate change doesn't seem, to the climate contrarians, like a wager at all, because they're convinced that mainstream climate science is bunk. Doing nothing isn't betting against the odds—as if one is putting money down on the wild chance of dodging catastrophe. Instead, it's entirely the sensible choice, because it's the one most likely to produce the best future outcome.

Of course, most contrarians, even ardent ones, don't explicitly say—likely even to themselves—that climate scientists are colluding liars. After all, the position has a basic credibility problem: unless one is ready to raise questions about the integrity of science generally—and few people are—then one must be able to explain why *climate* scientists, in particular, are so corrupt, while other groups of scientists, including those who receive large government and corporate grants for their research, aren't. Any such explanation is likely to sound transparently contrived.

When I've suggested to contrarian interlocutors that their sweeping disagreement with climate scientists' conclusions makes sense only if they're convinced, underneath it all, that climate scientists are collectively dishonest and engaged in a worldwide conspiracy, they almost always deny believing so. But in our conversations they'll often go on to argue why, fundamentally, that assertion is indeed true.

And look where we've arrived! Once again, the final and impregnable line of defense in a heated argument is that the other side—whichever one—is either stupid or (in this case) deceitful. And the impulse to deny that climate change might be catastrophic is so strong that climate contrarians are willing to impugn the honesty of *scientists*—professionals who are otherwise widely seen in Western societies as exemplars of objectivity and integrity.[21] Something very odd is happening here.

A second thought experiment helped me discover what that might be.

### MIND TRANSPLANT

In this case, I used a thought experiment to explore my own worldview—so it involved quite painful introspection. I sometimes felt as if I were conducting a mind transplant on myself.

I started by imagining what I'd think and how I'd feel if I were to learn that climate change is *not* a critical problem. Let's say some new scientific findings showed—beyond any reasonable doubt—that carbon dioxide is not a greenhouse gas and that the warming we've observed in recent decades is just part of a natural cycle that cooling will cancel out in time (as many climate contrarians do believe). How would I react to the news?

This thought experiment is, essentially, the reverse of the earlier one. Rather than holding constant key information from climate science and then seeing how we'd need to alter our worldview to justify inaction, we instead hold constant our worldview and then alter information from climate science so that action is no longer necessary.

Try it. It's exceptionally hard to imagine being in an alternative reality that contradicts a key belief. I found I needed to trace the countless implications of this new "fact" as they ramified in every direction through my network of concepts, beliefs, and values. But once I'd mastered the mental flip from believing climate change is a critical problem to believing it's not, I found I had two reactions in quick succession. Both involved strong emotions, but while the first was more rational and easily accessible to my consciousness—and could even be described as joyful—the second was more intuitive, harder to grasp and, frankly, much darker.

My primary reaction—by far the most powerful—was a feeling of enormous relief, even elation, when I realized my kids, Ben and Kate, wouldn't be endangered by climate change later in their lives. If humanity isn't causing catastrophic climate change, and cooling will come as part of a natural cycle, much of my worry about their future vanishes.

But my secondary reaction was much more surprising—it was something akin to subtle disappointment. Where did *that* reaction come from? At first, I couldn't tell. After a while, I identified one source: if climate change isn't happening, then a project around which I've built a large part of my life in recent years—the effort to rally people to address the problem—is suddenly pointless. That realization evoked a strong sense of disorientation and loss, partly for the time I'd apparently wasted fighting a non-problem, but more significantly because the project has brought great meaning to my life.

But that wasn't all. Holding the feeling of disappointment longer in my mind, I scanned farther down. And then one phrase popped to the surface: "Greed wins!" And then, "Selfishness and recklessness win!"

Bingo. I'd felt I'd seen a painful truth: at some level I'd come to *want* to believe that climate change is a particularly critical problem, because that reality aligned with assumptions, buried in my worldview, about the moral arc of the universe—about who the good guys are and who the bad guys are, and about what everyone should do to make the world a better place and what should happen to them if they don't. By this worldview, prudent action and care for our fellow human beings and for nature are morally good behaviors, while selfishness and reckless disregard for other people and nature are not just morally wrong because they hurt the innocent, but also irrational (in the sense, made famous by the philosopher Immanuel Kant, that such behavior can't be adopted by everyone without contradicting its underlying rationale). Some of the most selfish, reckless people are greedy owners and managers of big businesses, this latter view continues simplistically, and their actions are the main cause of climate change. Indeed, climate change is stark evidence of their behavior's depravity and stupidity. Yet if climate change didn't exist, then these people would "win" in the sense that they'd get away, once again, with their unbridled recklessness, imposing huge costs on others without any final moral reckoning. And deep down I found that possible future deplorable.

A little biographical background helps explain this reaction. On two occasions when I was younger—once in my late teens and once in my early twenties—I made a decision to pursue academic study rather than go into business. The second time I had a small amount of capital accumulated—some, ironically, from working on oil and natural gas rigs in Western Canada—and had reason to believe I'd do well in a couple of ventures in Canada's oil patch. I faced, fundamentally, that old choice between making money and learning more about the broader world. I chose the latter because I was alarmed by the state of international affairs at the time—chiefly by the risk of nuclear war between the United States and the Soviet Union. I went off to university to study why countries go to war and, more basically, why people fight each other.

I recall vividly the moment I took my first step along this path. It was July 1978, and I'd been working as a "motorman" on a gas rig, responsible for the massive diesel engines that drove the rig's drilling apparatus, winches, and pumps. We'd completed a well on the broad prairie, and the half-dozen members of the crew were taking down the derrick, a huge job in itself. I knew this was likely the last day I'd ever work on a rig. While I'd done the job for years, it wasn't one I relished. So, at the end of the shift, after the rig had been dismantled and loaded onto flatbed trucks, I threw my hard hat far into the air in jubilation.

Still, I've often wondered if I made the right choice. To sustain a key life decision for decades in the face of second guesses, it must be backed by strong reasons. Over the years my reasons became woven together in the deep fabric of my worldview as an interpretation of business—and of capitalist enterprise more generally—as a fundamentally selfish, shortsighted, and therefore often reckless endeavor that's motivated mainly by greed. Ever since my childhood, selfishness and imprudence have incensed me. So by associating business with these character flaws, I felt better about my decision.

But whenever this view popped into my conscious awareness, I recognized rationally that it is, at best, a crude stereotype. It might harbor a kernel of truth: we can all point to examples of capitalist venality—just take the behavior of many US bankers, hedge fund managers, and other financial bigwigs in the lead-up to the 2008 financial crisis, and after. But there's huge variation in the forms of capitalism, the behavior of businesses, and the character of individual business owners and managers. For every selfish creep, there are many who use business entrepreneurship to make the world better in some way. Also, selfishness can be a good thing. Adam Smith was right that the individual's pursuit of self-interest can, when directed through a well-functioning market, hugely benefit broader society. Modern market economies are prodigiously creative; and that creativity, when pointed in the right direction, can be an extraordinary force for good.

Most of the time, I remain largely unaware of my worldview's simplistic assumptions about the moral character of business and capitalism. But my second thought experiment brought them forcefully to my attention and showed me that my motivations to fight climate change and my views on who or what is responsible for the crisis aren't a straightforward result of my objective assessment of the problem. They're influenced, at least to some degree, by my desires to protect my sense of purpose and place in my moral universe—both fundamentally important to my self-esteem.

If so, how is my behavior different from that of my contrarian interlocutors? Could they be right that climate change is nothing more than a convenient and ultimately self-deceiving vehicle for my private psychological and ideological agenda? Actually, no, because at the center of my worldview, too, is the norm that decisions should be grounded in reason, empirical evidence, and science. If I were indeed confronted with scientifically grounded evidence showing, beyond a reasonable doubt, that humanity hasn't caused observed warming, that our emissions don't actually contribute to global heating, I believe I'd accept that evidence no matter what the implications for my self-esteem.

Right now, the scientific evidence that humanity is causing devastating climate change is stronger than ever. Nevertheless, after running through this second thought experiment, I'm much less inclined to attack contrarians with self-righteous zeal. I'm now far more aware how my own worldview—and the deeper, personal interests and motivations that it reflects—can subtly influence what I think is important and how I interpret events around me.

I'd recommend this kind of mental exercise if you want to understand better how your worldview affects your perception of the world around you, and especially how it supports your sense of purpose and self-esteem. Take a scientific, social, or political "fact" that's central to your worldview and that you regard as incontrovertible—say, a belief about intrinsic human equality or inequality, about the truth or falsity of a scientific theory such as evolution, about the reality of human agency, or (here's a big one) about the moral superiority of the national, religious, or ethnic group that's most vital to your personal identity—and see what happens when you tell yourself that this fact is simply not true. I expect you'll find it a somewhat wrenching yet fascinating exercise.

Psychologists say that we all unconsciously downplay evidence that doesn't support or justify our personal interests and motivations while highlighting that which does. They call this process "motivated

inference." As my colleague the philosopher and cognitive scientist Paul Thagard writes, such cognition is different from wishful thinking, because in this case our motivations don't directly produce our beliefs. "Rather, our goals lead us to acquire and consider information selectively, so that we manage to find some evidence that makes us think we are being reasonable in maintaining an emotion-based belief that we ought to doubt."[22] I've always regarded the term "motivated inference" as a euphemism for lying to oneself, and I've always believed it was a label for the other person's behavior, not mine. But the second thought experiment showed me that the little, motivated half-truths one tells oneself—in my case, about who's responsible for climate change—are exceptionally hard for one to see. Of course, that's why the scientific method—the process of bringing hard empirical evidence relentlessly to bear on our hypotheses and theories—is so important. Science isn't infallible, but it serves as a major check on such biased inference.

Yet, we're still left with a last, big question: Why do our worldviews have such power? Why do we grasp them so tenaciously that we're often willing to subtly twist facts and reason to fit into them and, even worse, impugn the integrity of others who see things differently?

Once again, my kids helped me find the answer.

# 16

# Hero Stories

*Man cannot endure his own littleness unless he can translate it into meaningfulness on the largest possible level.* Ernest Becker

AS BEN AND KATE HAVE grown up, I've been struck—and sometimes saddened and alarmed, as so many parents are—by their struggles to come to terms with death. I knew abstractly that an appreciation of death would dawn in them at some point, but I hadn't remotely expected the process would be so explicit and heartrending.

At around four or five years of age, a child starts to grasp, dimly at first, that death involves a kind of imponderable loss, that this loss is final, and that loved ones—mothers, fathers, grandparents—will eventually die. In our family the process has involved, indeed it still involves, age-appropriate, frank conversations with our children; and I've learned that the process through which children recognize and accept death's reality is enormously traumatic and that their emerging worldviews play a critical role in how they manage this trauma.

Ben asked his first explicit question about death when he was five and a half. He was having a bath and, with soap suds up to his chin, he paused his splashing and asked Sarah what happens after someone dies.

While Sarah and I have our individual convictions, we don't presume to have final answers to such questions. So she replied that some folks think that people who die go to heaven, while others think they're reborn as another person or animal or plant, while still others think nothing happens at all—that death is simply the end of life. Ben wasn't happy with the last option. "I think they go to heaven," he said firmly. Sarah offered that he might be right.

She told me about the conversation later, and we didn't think about it further. But the next day, Ben drew one of his remarkable pictures. On a piece of blue construction paper, he sketched a little boy dressed in an orange shirt and green pants. The boy's face was in profile, so he faced sideways across the sheet of paper. From a jagged and dark black slash representing the boy's mouth emerged a speech bubble containing the words "MUM DAD" in big, bold letters.

Sarah asked Ben what the picture represented. "This is me in heaven, Mum. It's blue paper because heaven is in the sky. And I'm shouting for you and dad, because I'm lost."

Kate started to ponder death a little earlier in her life. Her perplexity and discomfort revealed itself in ways different from Ben's. She started to ask the question "Will it die?" about things around her, both animate and inanimate—about trees, worms, rocks, and toads. Still, I remained largely unaware of her emerging concern about death till one morning when she was upset because something wasn't going quite her way. I put her on my knee and held her close:

"Sometimes you can't get what you want," I said. "That's part of what you learn as you grow up."

"But I don't want to grow up," she replied adamantly.

"Why?"

"Because then I'll have to die."

This was one of those moments—not uncommon as a parent—when one is at a loss for words. Kate's logic seemed unchallengeable: if she stayed a child, she could ignore the possibility of death.

I've found our children's struggles with the idea of death acutely poignant, but also enlightening: they've helped me see that fear of death is one of our most powerful motivators. We work hard to manage this fear and, in the process, often accomplish extraordinary things, many wonderful, yet some dreadful.

I think we fear death for five main reasons. Most obviously, we fear the physical discomfort and pain that often accompanies dying. People will sometimes say that they're afraid of dying but not so much of death. I agree we're afraid of dying, but I'm sure almost all of us are frightened of death too, whether we admit it or not.

We're frightened partly because we don't want to be separated from our friends and loved ones, especially from loved ones who depend on us. More metaphysically, we're frightened because death tells us we're transient and ephemeral; if we don't believe in an afterlife or a soul, then we likely believe that death will erase our consciousness—the very seat of our sense of self. Also, we fear that death will render our existence meaningless—that after all the Sturm und Drang of our lives, they'll have had, in the grand scheme of things, no reason, point, or purpose. We may even be unremembered, or at least unremembered as we'd like.

And we're frightened, perhaps most importantly, because death is unfathomable, and unfathomable things are usually scary. Absent a spiritual doctrine that specifically explains it, death remains one of the deepest, darkest unknowns of life. It's the ultimate, final edge between the known and unknown. The seventeenth-century English poet John Dryden put it perfectly in his epic drama, *Aureng-Zebe*:

> Distrust and darkness of a future state,
> Make poor mankind so fearful of their fate.
> Death in itself is nothing, but we fear,
> To be we know not what, we know not where.[1]

This is a noxious stew of fears. But considering death's emotional and metaphysical import, we give it remarkably little conscious thought. In wealthy societies, we can easily avoid thinking about death until a certain age, because it's mainly hidden away in hospitals and old-age homes. As we get into our fifties and sixties, we might ponder it a bit more, because friends start to die, as well as parents and older relatives, and because the remaining years seem suddenly so few. Still, we're largely unaware that we're always managing at one level or another the anxiety the prospect of death causes us.

### IMMORTALITY PROJECTS

The issue has been the focus of some fascinating research. In the 1960s and 1970s, the cultural anthropologist Ernest Becker drew on the work of Sigmund Freud, Søren Kierkegaard, Otto Rank, John Dewey, and other thinkers to develop a powerful theory of people's psychological relationship with the idea of death.[2] Since then, social psychologists Sheldon Solomon, Jeff Greenberg, and Tom Pyszczynski have elaborated Becker's idea into a framework called Terror Management Theory, or TMT; these researchers have then extensively tested their theory in the laboratory and the field.[3]

TMT proposes that fear of death is a primordial feature of the human mind. As early humans evolved their extraordinary intelligence, they started to use symbols in their minds to represent abstract ideas, like the ideas of the self and the future. Then, as we saw in chapter 6, they combined these two symbols in particular to imagine the self *in* the future. This recursive self-awareness—consciousness of one's consciousness through time—was a source of both awe and dread: awe, because it helped give these early humans a sense of agency and, with it, feelings of power and possibility; and dread, because it made them aware that their selves were subject to the inexorable degradation of time. "The most fateful consequence of mental time travel," the psychologist Michael Corballis writes, "may be the understanding that we will all die."[4]

Virtually all living things have a deep and ineradicable drive to survive. For early humans, the tension between that drive and the emerging awareness of eventual annihilation produced a focal anxiety, TMT researchers suggest. Realization that death can occur—randomly and uncontrollably—at any moment made this anxiety even worse. Today, it remains an intrinsic, though often subconscious, feature of our modern minds; and if we don't mitigate it in some way, it can overwhelm and paralyze us.

Of course, human beings have many motivations, and different circumstances can trigger different motivations in various combinations. But the need to cope with the fear of death is common to almost all of us, and powerful. So we tell ourselves stories about who we are and how we should act that help us believe we can heroically transcend death.

Such "hero stories" are infinite in their variety. We might, for instance, weave our concept of our self into a story about someone or something we see as noble, powerful, and enduring. In this kind of story, we could be a devotee of a god, follower of a charismatic leader, a member of a strong ethnic group or nation, or even a fan of a notable sports team. Or, we might instead tell a story in which we play the role of hero directly. Here, we could be raising a child, founding a company, fighting a war, discovering a new scientific fact, constructing a building, writing a book, leading a community group, saving the world—or perhaps just being a terrific employee or friend.

Whatever our hero story's content, though, it helps us believe that we're involved in an "immortality project," to use Ernest Becker's term: we foster literal immortality via the heavens, souls, and afterlives central to most religions, or allow a symbolic part of our self to persist beyond our physical death, in something like a child, company, building, book, ethnic group, nation, or a friend's memories. We hope and believe, Becker writes, that the things we create are of such "lasting worth and meaning" that they "outlive or outshine death and decay."[5]

So far, so good. But we can't create these hero stories and immortality projects out of thin air. To be compelling—not just to ourselves, but especially to the people who matter to us—they must connect with our surrounding culture. In other words, our hero stories need to make sense within the common worldview that we share with members of our group or community, whether it's a neighborhood association, sports club, political party, company, or nation. A group's worldview always includes a rich conception of "we," especially regarding why the group exists and its history; it also usually includes ideas about what counts as virtuous behavior and as fairness and justice in members' dealings with each other.

And here's the key point. When each of us wants to figure out what life projects will help our individual selves endure, literally or symbolically, we look to what's meaningful and valuable within the groups that we're members of and that matter to us. And to find out what's meaningful and valuable, we use as reference points these groups' shared beliefs about identity, good behavior, fairness, and justice. The groups we're part of are "codified hero systems," to use Becker's words. We seek to be heroes according to their codes; doing so gives us a sense of purpose and lessens our fear of transience.

Religion, of course, often serves this purpose. Regardless of the truth of any specific religious worldview, one of religion's psychological functions—perhaps its central function—is to alleviate its adherents' death anxiety. The stories that religions tell give their believers' lives meaning, make those believers' eventual deaths appear understandable, and often promise immortality through an enduring soul or spirit.

More generally, and maybe a bit cynically, one might say that we've learned to manage death anxiety by developing a prodigious capacity for denial. Some scholars even argue that denial of death is the secret of our species' evolutionary success.[6] Once we evolved this cognitive ability, or at least the ability to keep death anxiety at bay, we could deploy our intelligence, including our recursive consciousness, to further our survival and propagation—and eventually to dominate the planet—without suffering the paralyzing anxiety that this intelligence would otherwise have caused us.

As we try to craft an idea of astute hope in an increasingly dangerous world, we must recognize that the tension between our awareness of our certain death and our drive to survive is an immutable feature of our condition. It's one of those pitiless, inescapable aspects of our reality that we must accept and learn to work with. Death anxiety pushes its way into our consciousness when we're children, and it remains with us through our lives, although we do our best to bury it underneath our daily busyness. Some of us may "sublimate" it, as psychologists say, into a drive to do something exceptional or even noble. But even if we don't or can't, for our mental and social health, each of us still needs a hero story and immortality project, and our projects and stories must make sense within the shared worldviews of the groups that matter to us.

### EACH OTHER'S VILLAIN

When Ben at the age of six drew a picture of a submarine that could save the world's sharks, he was writing the first lines of a hero story. It drew on our family's larger worldview, with its keen awareness of nature, and on what seemed to be his innate moral impulses to prevent suffering and promote fairness. It gave him a purpose that addressed the problem at hand, and in doing so may have also alleviated some of his emerging death anxiety.

Over the next years, as Ben grew into a thoughtful boy, his stories became more detailed and emotionally variegated. By age nine he wanted to be an oceanographer—and to find Atlantis. His stories will continue to change as he gets older; it takes experimentation to find one that works, and we continually revise them as our lives evolve.

The hero stories we tell ourselves as adults are far more elaborate than those of children, because they must connect with the more elaborate network of concepts, beliefs, and values that make up our adult worldviews and the worldviews of the groups we're part of. They're also usually far less grandiose and narcissistic, because they must make sense in the context of hard realities—Atlantis, after all, doesn't exist—and because other members of our community generally find narcissism offensive. And, lastly, they're far less accessible to us in our everyday thoughts, because we feel a bit foolish when we think of ourselves as heroes and somehow immortal.

But the stories are still there, deep down. My story seems to be about being a good father, husband, teacher, and member of my community—yes—but also about reducing conflict in the world, calling out greed, selfishness, and recklessness, and protecting nature from human avarice and folly. These commitments are partly derived from the ideas of justice that I share with the liberal and progressive groups in Canadian society that I feel I'm part of.

And the climate contrarians I engage in conversation have their own hero stories. They too are members of families and communities—teachers, farmers, line workers, businesspeople, and others who sincerely believe that their hard work and enterprise are meeting people's needs and solving society's problems. For those who work in the fossil-fuel energy industry, in places like the oils sands in northern Alberta (people I know to some extent from my years working in the oil patch), digging up bitumen gunk and converting it into fuel for our cars isn't a hideous despoliation of nature but an exciting and noble expression of human will, exuberance, and self-determination. Many see themselves as challenging raw nature at the frontier of the Canadian wilderness and turning it into something enormously useful for everyday people. Key elements of their hero stories are moral commitments to personal freedom and responsibility and to the right to keep the fruits of one's ingenuity and effort—commitments partly derived from the notions of fairness and justice shared with their own (usually more conservative) groups in Canadian society.

When I pop up in their lives and start talking about climate change, they see someone representative of larger forces that could take away the fruits of their enterprise and limit their freedom. In their minds, I'm suggesting that government should steal their wealth and bind them in a web of rules and regulations, because it's hard to imagine any meaningful response to climate change that doesn't involve bigger and more intrusive government—and not just national government, but global government too. So I'm threatening to cut the heart out of their hero stories, the stories that protect them from the universal, omnipresent, and potentially overpowering fear of death and meaninglessness.

The commonly proposed solutions to climate change represent just about everything they dread most: constraint, impoverishment, and subservience—perhaps even subservience to foreigners. No wonder they get angry, and no wonder they're willing to do anything necessary—including dismiss clear scientific evidence as nonsense and even declare scientists to be liars—to defend themselves, their worldviews, and everything those worldviews mean to them.

On my side, when my interlocutors dismiss scientific fact and attack scientists, they threaten to cut the heart out of my own hero story, as I fight greed and recklessness to protect a planet I fear will die. I'm then inclined to see their moral commitments to personal freedom, private property, and the right to self-betterment as nothing more than a cover for rank selfishness—and for pillaging nature while the pillaging is good. Their expression of these commitments, by this view, simply confirms that they're reprehensible.

That's the essence of the mirror image: each side plays the villain in the other's hero story. Those villains and our individual ideas of justice make our heroism possible; we all see ourselves, we *must* see ourselves, as struggling for the good against the bad, even if we don't consciously admit it. And this kind of self-perception appears all the way up and down the socio-economic and power hierarchy on each side of the dispute, from workers angered by the burden of carbon taxes, to middle-class folks arguing about climate change around a dinner table, and on to American billionaires—such as Tom Steyer and Charles Koch—duking it out over climate policy by supporting opposing politicians in US congressional races.

So we go around in cycles of attack and counterattack, the debate becoming increasingly antagonistic and polarized, while climate change itself isn't effectively addressed.

Sensing this underlying psychological dynamic at work, some prominent commentators and academics—Ted Nordhaus of the Breakthrough Institute and Bjørn Lomborg of the Copenhagen Consensus are examples—have declared a "pox on both your houses." They've concluded that the truth about climate change must lie somewhere in the middle. The problem, they say, is neither as bad as the advocates for climate action claim nor as irrelevant as the contrarians assume.

But that's an error in logic. Just because each side exhibits roughly equal self-righteous fervor, it doesn't follow that each must be equally wrong about climate change. The social-psychodrama surrounding climate change tells us a lot about what makes us tick, but next to nothing about the underlying problem in dispute.

To learn about that, we need science, and the science says that the advocates for bold action are almost certainly right: climate change is a monumental threat to humanity's future.

## GETTING BEYOND FEAR

The mood shift that much of humanity has experienced in the last two decades—from excitement about the future's boundless possibility to deep pessimism about worsening insecurity and diminishing opportunity—still seems to be underway; it may even be gaining momentum in the wake of the pandemic. This shift is occurring, I'm convinced, because many of us, indeed perhaps most of us now, are increasingly afraid. And we're increasingly afraid largely because we can't reconcile the profound and rapid changes that we sense are happening around us with the assumptions about social order, fairness, opportunity, and identity that often remain at the core of our worldviews.

A problem like climate change is deeply contentious not just because of its complexity and likely severe consequences for people and societies down the road, but also because we know that if it's really happening, any meaningful response will implicate every facet of our lives and almost inevitably challenge some of our central worldview commitments, whether (for those of us on the ideological right) our commitments to limited government regulation and the unrestricted right to acquire wealth or (for those of us on the ideological left) our commitments to local, small-scale food and energy production and even to social equality and democracy, both of which will likely be ever-harder to sustain as the climate crisis worsens.

Our worldviews connect us with our communities, stabilize our sense of who we are as individuals and groups through time, anchor our visions of a desirable and hopeful future, and, not least, provide the raw materials for our personal hero stories. So we're terrified when they're threatened, and often come passionately and sometimes even blindly to their defense.

Quite understandably, some of us transform our fear into anger. Worse, rather than acknowledging that "the enemy is (partly) us" to explain the many disruptive changes we're experiencing, some of us create in our minds an external, personified enemy—an analogue of Sauron in *The Lord of the Rings*, under labels like *environmentalist* or *capitalist, white* or *brown*, or *Christian* or *Muslim*—whom we can blame for the disruptions and portray as our villains in new, angry hero stories.

In the end, though, as we all essentially know, such embittered reactions only make us more afraid and more divided—and collectively less able to solve our common problems. We need instead worldviews that are complementary enough to unite us around an immortality project for our entire species as we work to stop, and then reverse, the rapid deterioration of our planet's vital natural systems—worldviews that help us surmount fear by inspiring rather than extinguishing the hope that motivates our agency.

Humanity may have barely a decade or two to shift its dominant worldviews in such positive directions. To act so fast, we must understand better what's going on in our own worldviews and those of other people and groups. Then we'll see better who might be our natural allies, who might be persuaded to become our allies, and who's likely to oppose us implacably in the coming social and political battles for a better future.

# 17

# Strategic Intelligence

*If you know others and know yourself, you'll not be imperiled in a hundred battles.* Sun Tzu

WHEN ON AUGUST 31, 1961, the Soviet Union announced it was going to test nuclear weapons again in the atmosphere, breaking the moratorium it had observed with the United States and Britain since 1958, it declared that this time it was going to explode the largest bombs in history—in the range of fifty to one hundred megatons.

More than four years had passed since Stephanie May's first petition calling for an end to nuclear testing everywhere. In that time, she and her fellow anti-nuclear campaigners around the world had seen some progress towards a test ban. The most important step forward, by far, had been the moratorium, and they could reasonably believe they'd aided that progress. But with the Soviet Union planning to resume tests, Stephanie feared that everything they'd achieved would be lost.

Relations between the Soviet Union and the United States had been deteriorating since May 1960, when the Soviets shot down a US spy plane in their airspace. They'd worsened with a failed summit between the countries' leaders in Vienna in June 1961, and an escalating crisis over Berlin's wall-divided status.

On September 12, Stephanie went to New York for a board meeting of the National Committee for a Sane Nuclear Policy, or SANE. "It was a grim, heartbreaking gathering," she said.[1] Discussion focused exclusively on the Soviet decision. Board members were particularly upset that the media kept saying SANE wasn't against the Soviet tests—reflecting the widespread politically conservative assumption that peace groups were Soviet sympathizers—even though the organization had immediately declared its opposition to those tests and prominent SANE members had protested outside the Soviet mission to the United Nations in Manhattan.

To refute the dominant media perception, Stephanie proposed that SANE's entire national board start a hunger strike outside the Soviet mission, with members continuing individually or collectively until they couldn't physically go on. Almost the entire board reacted with patronizing incredulity. "What you're proposing is completely out of the question," one blustered. "SANE has an excellent public image as a responsible, dignified board of concerned, thinking people. . . . [This] is the worst suggestion of the day."

Stephanie went home to Connecticut, dispirited. Her mood worsened in the next days when a local newspaper published a vicious editorial cartoon attacking SANE for ignoring the impending Soviet tests.

> I was livid and more frustrated than I had ever been in my life. That night when I went to bed, I buried my head in my pillow and wept. John put his arms around me, kissing my tears, and begged me to stop crying. Finally, he said, "Tomorrow you go on your hunger strike. If no one else will do it, then you do it alone."

So, she did—alone. The next morning, she packed a sleeping bag, found someone to take over her carpooling duties, and asked the local chapter of SANE to issue a press release about her planned hunger strike. She arranged to have her two young children, Elizabeth and Geoffrey, stay with Norman Cousins and his family at their home outside New York. Editor of the renowned literary magazine *Saturday Review* and chairman of the SANE board, Cousins had also become a close family friend.

He was opposed to the hunger-strike idea; he thought her plan bordered on the absurd. She intended to walk back and forth in front of the Soviet mission all day long and then bed down on the sidewalk at night, with her small dog for protection. And because she wanted to remain at her "post," she was going to minimize her liquids consumption. Ellen Cousins, Norman's wife, was terribly concerned: "I've never heard of anyone going about a hunger strike that way. You could permanently ruin your health."

Despite his unease, Norman Cousins dropped Stephanie and her dog outside the Soviet mission on East Sixty-Seventh Street the next morning. Dressed in a tweed suit and three-inch spike heels, Stephanie introduced herself to the pair of armed patrolmen in front of the building, telling them why she was there. They immediately threatened to arrest her as a vagrant. Soon, their captain arrived in a police car. Because the sidewalk was considered Soviet territory, he told Stephanie, she'd need to keep walking all day long so that she could be considered a pedestrian. He also told her to leave by 6 p.m. and not arrive till 8 a.m. the next morning. She agreed to all these conditions and, once the captain had left, unfurled her sign reading "RUSSIA! STOP Nuclear Testing!! Stop poisoning the air!" She tied it around her neck and began walking.

She remained at her post—from eight in the morning until six in the evening—for nearly six full days, sleeping at a wretched local hotel and drinking only a few cups of juice, coffee, or broth a day. After the first day, sore feet, dizziness, and harassment by the police forced her across the street. A nun at an adjacent parochial school refused to give her a chair. Passersby glanced away, while others told her to get out of the neighborhood. But mostly, she said later, "few people seemed to notice me at all."

> Yet there I was, making an absolute spectacle of myself, walking around with a sign that stretched from my head to my feet, and I was absolutely invisible, except to little children instructed not to look, and to teenagers in parochial school uniforms who stole furtive glances and then giggled.

The situation didn't get much better for days. A few friends dropped by to offer support (one supplying a sign showing how many days she'd been on the hunger strike), and a kind woman in a nearby apartment provided a chair. But she was nearly evicted from her hotel, and her crushing invisibility persisted, except when people's anger burst to the surface. On the fourth day, one of the patrolmen in front of the Soviet mission crossed the street to confront her. "Listen," he said, "I was in Korea. I fought the Commies then, and I'm ready to fight them now. As far as I'm concerned, you and that so-called sane committee you're tied in with are just aiding and abetting the enemy, with all your ban-the-bomb commie-rot slogans. You're all either crazy or pinko, and you make me sick. So, don't expect any favors from me!"

But under the surface, almost imperceptibly, events had started to tip her way. A few local reporters dropped by, and a couple of small stories appeared in the New York tabloids. The Associated Press wired nationwide a photo of Stephanie holding her sign as she sat in her chair; the caption said she was a member of the national board of SANE, so recognizing the organization's opposition to Soviet tests.

That night a reporter from the *New York Post* called her hotel room at 2 a.m. "I expected some sort of fanatic," he confessed after an extended chat. "You sound not only very reasonable, but extremely articulate, especially for this hour of the day." Stephanie assured him that she was indeed a fanatic. The next afternoon the *Post* published a long, favorable story about her protest. The piece quoted her as saying: "I'm not willing to crawl into a hole in the ground and accept nuclear destruction without a murmur. This is my murmur."

On Friday morning, John joined her on the sidewalk holding one of the signs. By early afternoon he remarked that it was enervating to have "so many people just walk by without looking or saying anything." But a moment later a panel van from ABC television pulled up and disgorged onto the sidewalk reporters, cameramen, cameras, cables, batteries, and audio equipment. They'd come to interview Stephanie. As soon as the ABC crew had left, one from CBS arrived, and over the next several hours, filming vans from NBC, Universal Pictures and Movietone, and New York's main local television channel pulled up, too, spreading their filming paraphernalia across the sidewalk.

Despite her exhaustion and weakness, Stephanie was ready for this moment. Why was she there? the interviewers all asked. Did she think her hunger strike would accomplish anything? How could it have any effect on the Russians?

She took the opportunity to give each interviewer a quick synopsis of the dangers of nuclear fallout, especially to children. She acknowledged that her protest wouldn't have any direct effect on the Soviet leadership, but she said she'd heard that people in the mission were very unhappy about her protest. "I feel that makes it worthwhile; they're doing a terrible thing, poisoning the atmosphere, and they

should be unhappy." A "public outcry throughout the world" could stop the crime of weapons tests. And once again she happily admitted she was a fanatic. "In the nuclear arms race, there are no winners," she explained, "and the finish line at the end of the race is just that, the finish line—the end of the human race, the end of everything." She had, she said, "a fanatical desire to stop that from happening."

In the evening, when they had to vacate the sidewalk, Stephanie and John went to a SANE fundraising event. Television sets had been placed around the hall, each tuned to a different channel. All three national broadcasters' evening newscasts featured the protest, mentioning that Stephanie was a SANE board member. The people clustered around the TV sets cheered. Board members who'd condescendingly dismissed Stephanie's idea now warmly shook her hand.

The next day, John convinced Stephanie to end the strike, pointing out that she'd accomplished more than she'd dreamt possible. But before leaving, they went into the mission to register their protest directly with the Soviet government. A member of the diplomatic core sat with them, taking notes that he said he'd pass on to Moscow. Then, as they stood up to go, he mentioned that he wanted nuclear testing to end too. "The arms race is in the interest of no country," he said, looking deeply sad. "There is no family in Russia untouched by the last war . . . we lost millions of people. I can't imagine anyone surviving a nuclear war."

In the weeks that followed, people across America began hunger strikes to stop nuclear testing.

### TRANSCENDING WORLDVIEWS

When Stephanie imagined a future situation that could be the object of her hope, whether that possible future was days or years away in time, her honesty tempered her expectations. It tempered her expectation of the influence she could have on the Soviets; it tempered her estimate of the possibility of achieving a nuclear-free world; and it tempered—too much, it turned out—her expectation of the impact she could have on public opinion, by herself, through a hunger strike. Before she began, she imagined she'd have a large effect only if many other mothers joined her on the sidewalk. It never occurred to her that she'd have great impact simply by sitting there alone—that her determined, noble isolation would so intrigue the media and strike a chord nationwide.

She was also astute. She understood better than her colleagues the importance of the evocative act that could break through people's hardened perceptions. And she understood better than her world-wise counterparts that, given the public's anti-communist antipathy towards peace groups, any act by SANE as an organization would have to be truly dramatic, and the cost to her as a person meaningful, for people to rally behind it.

Stephanie understood these things because she was adept at comprehending other people's worldviews—even those she fundamentally opposed—and then using that comprehension to work with or around people to reach her goal. Of her many formidable strengths, this was perhaps her greatest, and she brought it to bear repeatedly in her activist campaigns throughout her life. Whether she was dealing with skeptical politicians in Washington, DC, or a local newspaper editor—first adamantly opposed to her campaign and eventually one of her biggest supporters—she started any interaction by seeking to understand her interlocutor or her opponent. Yet she was far less the zealous missionary aiming to convince people of the rightness of her position, and far more a mobilizer aiming to rally people already convinced, or at least concerned, but feeling too disorganized, impotent, or hopeless to act.

Stephanie is, in fact, a case study of the virtuous circle of hope and agency that I wrote about in chapter 5. Because she worked hard to understand people's fears and passions, their strongest motivations and weakest points, she was more effective in her various projects. Her effectiveness then reinforced her hope and encouraged her to keep pushing. It turned a passive, wistful attitude—a "hope that"—into a motivating "hope to" and a potent instrument of political strategy.

People of course constantly underestimated her, even her friends and allies, but most significantly her opponents, who patronized her as a mere housewife. Evidence of that attitude appears over and over again in the letters people sent her, now pasted in her old scrapbooks. Friends and opponents thought she'd eventually fold up and go home, given the pressure she was under and the obstacles she faced. But she knew that they hadn't reckoned with the motivating power of her hope. It was her secret trump card. Whatever power her hope lost because it was honest, it more than recouped because it was astute.

In short, she had an instinct for leverage points in her relations with others. Remember that the environmentalist and systems theorist Donella Meadows ranked "the power to transcend paradigms" as our most effective lever—more than, for instance, changing a system's feedbacks or information flows. Meadows argued that the concepts, beliefs, and values that make up our paradigms, or worldviews, are the foundations on which all human social and technological systems are constructed. But she didn't put paradigms themselves at the top of her effectiveness list (the table on page 241)—she ranked them number two. Instead, in a truly brilliant observation, she reserved that special place for the ability to *step outside* paradigms—to, as she put it, "keep oneself unattached in the arena of paradigms, to stay flexible, to realize that *no* paradigm is true." This ability provides a "basis for radical empowerment," because it enables us to see vastly more options as to what we might do and where we might go. "It is in the space of mastery over paradigms," Meadows concluded, "that people throw off addictions, live in constant joy, bring down empires, found religions, get locked up or 'disappeared' or shot, and have impacts that last for millennia."[2]

When Stephanie stepped outside her own paradigm to make sense in her mind of other people's worldviews, she was using the natural human capacity for recursive thinking—she was thinking about other people's thinking. This kind of self-reflection fills her memoirs. It made her more *strategically* effective, and not just because it helped her better understand other people's passions, motivations, and weaknesses; it also helped her see how other people saw her. This gave her some distance from her actions, allowing her to judge their effectiveness more objectively and, in turn, better strategize her "moves" in the political game she was playing. As the famous sociologist Erving Goffman once noted, one makes such moves "in the light of one's thoughts about the others' thoughts about oneself."[3] Strategy involves mental gymnastics such as "I believe that you believe that I'll do A, so instead I'll do B."

Some might object that Donella Meadows's recommendation that we cultivate the ability to detach ourselves from our worldviews—"to stay flexible, to realize that *no* paradigm is true"—is a direct route to post-truth muddleheadedness at best, and social chaos at worst. Aren't some worldviews more "true"—more accurate reflections of the world's underlying reality—than others? Undoubtedly, yes. But no worldview is true in any final sense. And many aspects of our worldviews are matters of judgment and value, about which there is, ultimately, no final truth. Further, to see the world as others see it, and to be strategically smart in one's interaction with those others, one needs to start by accepting, at least temporarily, the things they believe to be true, much as I did when I accepted temporarily the truth of the nonexistence of climate change, for the sake of trying to understand a radically different point of view.

I mentioned earlier that, in the past decade, my colleagues at the University of Waterloo and I have developed two tools—the state-space model of ideology and cognitive-affective maps—that can help us transcend worldviews in just the way Donella Meadows prescribes. They make it easier to see deep inside our own worldviews and, also, to engage more constructively with people guided by worldviews quite different from ours, so we can better solve common problems, identify and address disputes about fact, and build together a vision of a future worth having. These tools, I believe, can help efforts to reduce the discord and contention—the ideological, social, and political polarization—that now poisons so many of our societies and threatens to ruin any chance that we might work together successfully to address crises like climate change.

My colleagues and I have used these tools to explore the underlying thinking of people with whom we disagree vehemently. And in doing so, we've found ourselves less inclined to jump to conclusions—for instance, that such people are stupid or deceitful. We might still have reason to believe their viewpoints are empirically false or morally wrong in certain ways, but less often do we think these people are just crazy or liars. We can better move beyond befuddlement and anger.

The two tools can also enhance our strategic smarts, so we can better engage in political action and mobilization and, when necessary, more effectively circumvent or even defeat our political opponents. And finally, they can boost our imaginations to explore and better understand what I call the "mindscape"—the landscape of all possible worldviews available to human beings, perhaps to find new worldviews that people have never experienced before and that could help us prosper in the future.

# 18

# Mindscape

*Most people believe the mind to be a mirror, more or less accurately reflecting the world outside them, not realizing on the contrary that the mind is itself the principal element of creation.* Rabindranath Tagore

MY PART IN DEVELOPING the state-space tool began nearly forty years ago.

In 1982 and 1983, when I was in my mid-twenties, a friend and I took a long backpacking trip overseas. In those days, many backpackers voyaged roughly along the equator, flying between tropical countries like Indonesia, Thailand, and Kenya. We decided to go instead from north to south and to travel as much as possible overland. We began in Finland, ended in South Africa, and in between visited eight countries, including the Soviet Union, India, and Zimbabwe. We chose this route to encounter as wide a range as possible of political, economic, and social systems. And in those days, the differences were breathtaking.

Soviet-style communism seemed alive and well in the countries of Eastern Europe; India was trying to find a distinctly South Asian economic path; and apartheid was still vigorous in South Africa.

Capitalism was, of course, an enormously powerful force in the world, yet it remained largely a Western phenomenon. Although Chinese leader Deng Xiaoping had begun introducing market reforms, particularly in the agricultural sector, to unshackle his country's economy, the rest of the world hadn't recognized the staggering implications of his changes. And truly globalized capitalism—its power raw and relentless—hadn't yet hyper-accelerated the homogenization of humanity's worldviews, institutions, and technologies.

We were a grubby pair, curious and adventurous. We broke Soviet tourist rules to visit rural Ukraine and Uzbekistan, hitched rides on trucks and freight trains in Africa, and pulled strings to meet senior diplomats. We treated the trip like a rolling seminar, collecting and reading huge amounts of local materials along the way—history books, translated novels, political magazines, religious philosophy, national propaganda, and even government legislation—and then mailing much of it home in boxes.

We found the contrasts between the societies we visited sometimes bewildering beyond our imagining. It was one thing to learn from afar about Finnish socialism, Soviet authoritarianism, the Indian caste system, Nepalese Buddhism, Maasai communalism, and South African apartheid. It was quite another to see these social phenomena up close and to talk to people accepting—or, in some cases, opposing—the worldviews, institutions, and technologies underpinning them. My friend and I asked ourselves what made these societies so wildly different from each other. We puzzled over how human beings who are, as Abraham Maslow has argued, basically the same in their biological and psychological needs for food, shelter, sex, community, and identity could create, as their various societies evolved through time, such vastly diverse ways of living together. It's a puzzle raised in the opening paragraphs of any basic anthropology, sociology, or comparative politics textbook, of course, but its familiarity doesn't make it trivial. Brilliant people through history have struggled to address it, starting most famously with Aristotle, and it perplexes all but the most unreflective traveler.

One evening in the Soviet city of Volgograd, my friend and I were chatting over dinner. The city, called Tsaritsyn until 1925 and then Stalingrad up to 1961, was the site of a titanic clash between Axis and Soviet forces in 1942 and 1943—arguably the single most important battle of World War II. Earlier in the day we'd visited the battleground's remarkable memorials, including the towering figure of Mother Russia calling the nation's sons to her defense, located at the top of Mamayev Kurgan, a hill that was a critical objective in the battle.

We were wondering about the underlying clash of beliefs between the Nazis and the Stalinist Soviets, and more generally, about how differences in worldviews contributed to the remarkable diversity of societies and social behaviors we were seeing on our travels. Perhaps, we thought, these differences could be traced to people's different answers to some basic questions regarding the fundamental nature of the world around them, particularly the social characteristics of that world. On the back of a paper serviette, I scribbled down five questions I thought were key:

1. Are moral principles universal and objective?
2. Can people choose their fate?
3. Are there large and essential differences between groups of people?

4. How much should we care about other people?
5. Should one resist authority or defer to it?

The first question taps the age-old debate between moral absolutism and moral relativism; the second raises the question of the limits to our agency or free will; the third asks how much we think other groups of people are like us; the fourth asks about our sense of responsibility to others; and the fifth speaks to our willingness or unwillingness to accept the use of coercive power. I assumed that the answer to each question could be placed somewhere on a single scale, with diametrically opposed answers at each end, so I drew the scales like this:

Morality is:

relative ———————————————— absolute

A person's ability to choose his or her fate is:

low ———————————————— high

Differences between groups are:

small and unimportant ———————— large and essential

We should care about other people:

a lot ———————————————— not much

In our response to authority, we should generally:

resist ———————————————— defer

The scheme hugely simplified the way people think about the political, social, and moral issues affecting them and their societies (a component of our broader worldview that social scientists sometimes loosely call "political ideology"), and we recognized that these questions are in many ways unanswerable in any absolute sense.[1] Still, we agreed they're more or less inescapable: any group of people needs to answer most of them, at least in a rudimentary way, to develop a shared understanding of the nature of the surrounding world and of what counts as morally good or bad behavior in that world—both of which are central to a group's understanding of its identity and purpose.

And elementary as the scheme was, we saw that different combinations of answers would lead to radically distinct political perspectives.

For instance, if a group's members generally believe they should care a lot about other people, while also thinking that differences between groups are small and unimportant, they'll likely have a broad, communitarian worldview: their understanding of "we" will tend to encompass all people everywhere, so they'll be more likely to have feelings of responsibility for, and be prepared to help, even those far away.

But if the group's members combine that same belief in responsibility to others with the belief that differences between groups are large and essential, they'll still have a communitarian worldview, but their "we" will likely be far more exclusive—perhaps centered on a national, ethnic, racial, class, or religious identity that defines a clear line between "we" and "they." The Nazis, of course, adhered to a brutal racial identity of this kind, while the Stalinists fused Russian nationalism with proletarian class consciousness.

We arranged the answers to the questions to create stereotypically "left" and "right" political views. Combining the positions on the right end of each scale yields a worldview emphasizing moral absolutes, individual agency, essential differences among groups, responsibility mainly to oneself, and willingness to defer to authority. In the North American context, this is a fairly conventional right-wing perspective. Combining the positions on the left end of each scale yields a worldview emphasizing moral relativism, the power of circumstances over choice, the essential similarity of all people, responsibility to others, and resistance to authority—a common leftist perspective.

But a shift in a person's beliefs on any one of these dimensions can lead to a striking shift in perspective, we noticed. For example, flip the right-winger's attitude on authority—from believing one should generally defer to authority to believing one should generally resist it—and the conventional conservative turns into something resembling an American libertarian, such as a follower of the late US author and philosopher Ayn Rand. The left-winger's position on moral principles can be similarly flipped, because not all lefties lean towards moral relativism. Some are as adamant as conservatives about moral absolutes, but they take different principles to be absolute. While a conservative might emphasize the sanctity of life, for instance, someone on the left might stress the principle that women should control their own bodies (and therefore support abortion rights).

My friend and I were so absorbed with these ideas, we didn't notice that we were the last people in the restaurant. Eventually, one of the waiters suggested, with a certain Soviet-era brusqueness, that it was time for us to leave.

### STATE SPACE

When I began my academic career in Canada years later, I came across somewhat similar approaches in political science, cultural anthropology, and social psychology. But most of them, including my own, seemed too unsophisticated to explain the enormous range of human worldviews. So I put aside the puzzle for more than two decades to focus on how environmental stresses—shortages of fresh water, for example—can cause conflict and how societies can innovate to address these problems.

When in 2008, I moved to the University of Waterloo, the exciting research community I found there inspired me to return to the worldview puzzle. I dug around in an ancient filing cabinet in my basement to find my decades-old notes on the basic questions underlying political ideologies. At once, I realized that my list of five questions was woefully incomplete; there were certainly more. But how many of significance underlay the range of ideologies we see in the world, and what were they?

Our research group soon expanded internationally to include leading scholars on ideology in the United States and Europe, and together we studied previous research on the topic going back over a century, as well as reams of recent data on political attitudes within and across societies.[2] Drawing on an idea from complexity science, I suggested that we could use basic questions like those my friend and I had identified to create a "state space" that would help us visualize the diversity of political ideologies.

Here's how the idea works.

Let's assume that the ideologies we're studying answer only the first three of the questions I jotted down on the back of a napkin in Volgograd—those about morality, free will (agency), and group differences. Let's also assume that each question has only two answers, those given by the opposite ends of the question's scale. Then, if we arrange the three scales so they're perpendicular to each other, we create a cube—like the one shown here. This cube has eight smaller cubes inside it, each standing for one combination of answers to the three questions. The whole cube itself is the state space—it's the *space of all possible states* of political ideology in this imaginary, oversimplified world.

That's clear enough, perhaps even trivial; but here's where things get more interesting. After working with my colleagues, I revised my original list of five questions and added ten more—as shown in the table on page 306. I divided the questions, each addressing a specific "issue," into two broad categories: those concerning beliefs about

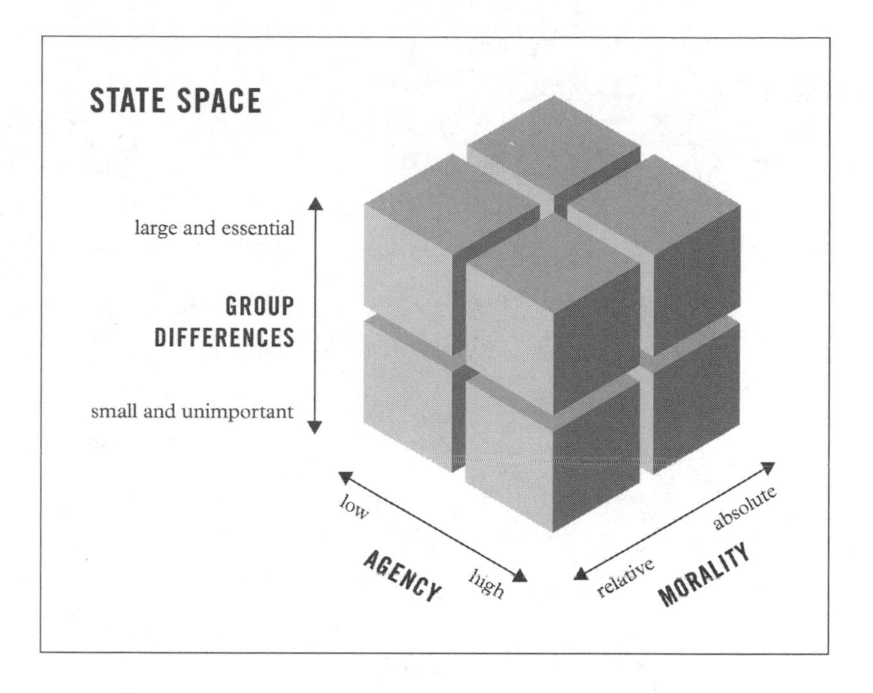


basic facts about the world (what I call "is" questions) and those concerning beliefs about how people should see or behave in that world ("ought" questions). This is/ought distinction is usually attributed to the eighteenth-century Scottish philosopher David Hume. Modern philosophers have shown that the line between the categories is fuzzy, but the distinction is still a useful way to order lists like this one.

Then, just as my friend and I had done in Volgograd, I posited two polar-opposite answers to each question in the table, placing one at each end of a scale. But this time, I arbitrarily divided the scale into five segments, with each segment standing for the strength of a person's belief in their answer to the specific question: strong ("S") and moderate ("M") on each side of a middle answer of "Ambivalent/No position." (You can find more information on the table and its scientific foundations at www.commandinghope.com).

**IDEOLOGICAL STATE-SPACE QUESTIONS**

| | ISSUE | QUESTION | BELIEF STRENGTH | | | | | | |
|---|---|---|---|---|---|---|---|---|---|
| IS | *Threat* | Is the world a safe or dangerous place? | SAFE | S | M | Ambivalent, no position | M | S | DANGEROUS |
| | *Source of Understanding* | Is the world best understood through reason or feeling (emotion/intuition)? | REASON | S | M | Ambivalent, no position | M | S | FEELING |
| | *Spirituality* | Is the world infused with a spirit? | MATERIAL | S | M | Ambivalent, no position | M | S | SPIRITUAL |
| | *Moral Principles* | Are moral principles subjective or objective? (Relativism vs. absolutism) | SUBJECTIVE, CONTEXTUAL | S | M | Ambivalent, no position | M | S | OBJECTIVE, UNIVERSAL |
| | *Agency* | Is a person's fate a result of circumstances or choice? (Determinism vs. free will) | CIRCUM-STANCES | S | M | Ambivalent, no position | M | S | CHOICE |
| | *Human Nature* | Are people basically generous or selfish? | GENEROUS | S | M | Ambivalent, no position | M | S | SELFISH |
| | *Relationship between Humans and Nature* | Are human beings as one with nature or distinct and exceptional? | AS ONE WITH NATURE | S | M | Ambivalent, no position | M | S | DISTINCT AND EXCEPTIONAL |
| | *Social Differentiation* | Are the differences between groups small and unimportant or large and essential? | SMALL AND UNIMPORTANT | S | M | Ambivalent, no position | M | S | LARGE AND ESSENTIAL |
| | *Source of Personal Identity* | Does one's identity derive mainly from oneself or from one's group? | FROM ONESELF | S | M | Ambivalent, no position | M | S | FROM ONE'S GROUP |
| OUGHT | *Time* | For inspiration, should one look to the future or to the past? | TO THE FUTURE | S | M | Ambivalent, no position | M | S | TO THE PAST |
| | *Change* | Should change be encouraged or resisted? | ENCOURAGED | S | M | Ambivalent, no position | M | S | RESISTED |
| | *Care for Others* | How much should one help others? | A LOT | S | M | Ambivalent, no position | M | S | NOT MUCH |
| | *Authority* | All things being equal, should one resist authority or defer to it? | RESIST | S | M | Ambivalent, no position | M | S | DEFER |
| | *Power* | Is the use of power over others usually wrong or often right? | USUALLY WRONG | S | M | Ambivalent, no position | M | S | OFTEN RIGHT |
| | *Wealth* | Are large differences in wealth immoral or moral? | IMMORAL | S | M | Ambivalent, no position | M | S | MORAL |

The state space now has fifteen dimensions—as before, one dimension for each question. Yet this time we can't really imagine the space as a physical object like a cube, because we live in a world of only three spatial dimensions. And in this case the number of answer combinations tucked inside the space is a lot more than eight—in fact, it's over thirty billion! (The number of answer combinations is equal to the number of answers per question to the power of the number of questions. Hence, two answers to each of three questions permit $2^3$ or eight possible combinations; two answers to each of five questions permit $2^5$ or thirty-two possible combinations; and five answers to each of fifteen questions permit $5^{15}$ or more than 30.5 billion possible combinations.)

In my mind's eye, I imagine stepping inside this multidimensional space and seeing an endless expanse of tiny dots, like the pixels on a computer screen, spreading out from me in every direction, with each dot representing a specific combination of answers to the fifteen questions.

In this gargantuan state space, people will prefer some broad zones of dots over others. For one thing, people's personalities and temperaments affect what political ideologies they adopt—and therefore how they'll answer the questions. People who are generally intolerant of ambiguity, anxious about loss, and unreceptive to new experiences, for instance, tend to have more conservative worldviews, probably because these views help them cope with everyday uncertainty and threats (and, underneath it all, perhaps ultimately, anxiety about death).[3] Since some kinds of personality and temperament are more common than others (lots of people are intolerant of ambiguity, for example), we'd expect the political ideologies these people prefer—and so the answers they give to the questions—to be more common too.

Also, research shows that people often share certain common moral intuitions, which leads them to prefer certain worldviews or political ideologies over others. After surveying a vast range of psychological, anthropological, and cultural research, the social and cultural psychologist Jonathan Haidt and his colleagues have concluded that six common moral intuitions—relating to care, fairness, loyalty, authority, sanctity, and liberty—affect people's ideological positions.[4] (Liberals tend to anchor their worldviews in intuitions about care, fairness, and liberty, while conservatives tend to draw on all six intuitions, the researchers found.) People will answer the state-space questions differently depending on what moral intuitions they find most compelling. For instance, someone committed to the moral value of protecting personal liberty is likely to say (in answer to the question in the table about free will or "Agency") that a person's fate is more a result of choice than circumstances.

## BASINS OF ATTRACTION

Metaphorically, the zones people prefer in the state space are a bit like valleys, or depressions, in a landscape—or "basins of attraction" in what complexity scientists call an "energy landscape." Just as water will flow into a depression in a real landscape on Earth and form a pool at that spot, people's beliefs will naturally gravitate to a comfortable basin in their worldview state space and settle there. Unstable worldviews or ideologies, on the other hand, are high points on the energy landscape. Just as it's hard to keep water at high points on a real landscape, people find it hard to sustain worldviews with configurations of belief that don't resonate with their personality types or moral intuitions. Their worldviews will then tend to migrate towards more coherent and appealing configurations, like water flowing downhill to ponds and lakes at a lower elevation.[5]

As I step into the state space in my mind's eye, I imagine that the fifteen-dimensional landscape of pixel dots has somehow been converted into our three-dimensional world and now resembles a terrain we might find on Earth, with a pixelated topography of mountains, valleys, plains, and cliffs. I call this place the "mindscape": it's the geography of all possible human worldviews.

Some might object that the wording of the fifteen questions is too beholden to mainstream Western culture and received history, given my social, economic, and intellectual background and that of my collaborators. Or, they might say that even if we can discern questions that are common across cultures, arraying each answer along a single scale between polar-opposite (dichotomous) answers is arbitrary and quite possibly misleading.

But most of the questions, research suggests, are indeed common across cultures. The American political psychologist Kevin Smith and his colleagues note that "all mass-scale social units face common dilemmas" of this sort.[6] I've heard similar questions asked in everyday conversations from China to Zambia, and I'm sure many of you have had comparable experiences in your own travels. Also, any broad explanatory device like this has to be arbitrary to some extent; some details and nuances must be sacrificed for clarity of insight. In this case, to create a state space, we must array each question's answers along a single dimension, with opposite answers at each end. And human beings do tend to think dichotomously; for instance, we like to see things as being either clearly inside or outside their relevant categories.

Still, taken together, the questions I've listed are best seen as a working hypothesis about key underpinnings of people's political ideologies and their worldviews more generally. My collaborators and I are refining and adjusting them as we learn more, and we'll publish our findings over coming years. But I've decided that the early results are strong enough—even though we haven't yet dotted every scholarly *i* or crossed every *T*—to warrant presenting the questions here so that people can begin to use them.[7]

They can help us in two vitally important ways.

First, by creating an imaginary yet well-defined space—the mindscape—in which most if not all existing political ideologies reside, they can help us see better where those ideologies sit relative to each other, especially how close or far apart they are. The spatial metaphor also makes it easier to view our own and other people's worldviews from alternative perspectives—and perhaps improve understanding between ourselves and other groups. "By recognizing the underlying structures of meaning instilled in us by our own culture, we can become mindful of our own patterns of thought," says the American author Jeremy Lent.[8] And by shifting our answers to the questions, we can imagine moving across the mindscape towards another worldview.

In just this way, for example, the questions have helped me see more accurately where the climate contrarians I sometimes debate are coming from. We know that their deep skepticism of government regulation is often rooted in strong commitments to personal agency and the moral right to accumulate wealth. So in my debates with them, I sometimes mentally visualize shifting my own views on these issues step by step, as if moving a slider along the agency and wealth scales in the table. And as I do, I feel like I'm walking towards them through the mindscape.

Second, we can also use the questions to help us imagine alternative worldviews. Together, the questions are a bit like a laboratory bench on which we can mix different combinations of answers—rather than different combinations of chemicals—perhaps creating some stable worldview alternatives that otherwise wouldn't have occurred to us.

Why do we need this kind of imagination boost? Because all societies tend to adopt two or three dominant political ideologies, or basins of attraction in the mindscape; then, as we've seen in recent decades in the West with modern conservatism and liberalism, they stay locked in a cycle between them for extended periods. Meanwhile, vast regions beyond those basins remain unseen and unexplored; and because we can't see, or because we resist exploring, any other basin—that is, other possible configurations of our beliefs and values—we become blind to even the possibility of alternatives.

When I traveled overseas with my friend in the early 1980s, the world's social and ideological systems were still markedly diverse. Of course, we found a few of them bizarre, and some morally reprehensible, but the experience still stretched our imaginations. Since that time, rising connectivity and a particular kind of globalized capitalism have homogenized many of humanity's worldviews, institutions, and technologies, creating, essentially, a vast basin of attraction around liberal-market ideas of economic individualism, property rights, limited regulation, and a consumerist notion of the good life.

Today, most of humanity's significant worldviews are either well inside this basin or orbiting around it. Many are inside, in the sense that they share the dominant capitalist ideology's basic tenets. Even the new populist authoritarianism that has arisen in societies as diverse as Hungary, the Philippines, Brazil, and the United States hasn't (yet) rejected capitalism's core principles. A few partially developed contemporary worldviews—like those of the Occupy protests of 2011 and 2012, the Transition Town movement promoting local economic autonomy, the youth climate strikes of 2019, and (as an example at the other extreme) the violent jihadist Islam that has arisen globally in the last thirty years—orbit around this dominant worldview, partly by defining themselves in opposition to aspects of it. Overall, then, humanity's conversation about alternative worldviews is astonishingly impoverished—and dangerously so, too, because we desperately need new ideas about how to live together on our imperiled planet and redefine our relationship with Earth's material environment.

One way we can enrich this conversation, and at the same time radically change our mindscape's topography, is to introduce questions that create new dimensions in the mindscape. For example, until relatively recently the worldviews of modern Western societies almost always represented human beings as fundamentally distinct from, usually superior to, and having mastery over the surrounding natural world. There was literally no question about this aspect of our reality. But in the last sixty years in the West, scientific findings from biology and ecology have combined with environmental activism and greater attention to Indigenous ideas to reintroduce into popular awareness the earlier, fundamental recognition—represented by my table's seventh question—that humanity is deeply enmeshed in nature and so intimately dependent on it. In 1963, less than two years after Stephanie May was protesting on the New York City sidewalk about the dangers of nuclear radiation, the pioneering biologist Rachel Carson, author of the seminal environmental clarion call, *Silent Spring*, wonderfully described the meaning of this shift in viewpoint:

> We still talk in terms of conquest. We still haven't become mature enough to think of ourselves as only a tiny part of a vast and incredible universe. Man's attitude toward nature is today critically important simply because we have now acquired a fateful power to alter and destroy nature.
>
> But man is a part of nature, and his war against nature is inevitably a war against himself. The rains have become an instrument to bring down from the atmosphere the deadly products of atomic explosions. Water, which is probably our most important natural resource, is now used and re-used with incredible recklessness.

> Now, I truly believe, that we in this generation, must come to terms with nature, and I think we're challenged as mankind has never been challenged before to prove our maturity and our mastery, not of nature, but of ourselves.[9]

That was nearly sixty years ago, and we've still not fully embraced the idea that humanity is part of nature, at great cost to all of us. But Rachel Carson's words show that the most vital innovations in our worldviews or political ideologies in our future could involve adding new dimensions to the mindscape, perhaps dropping current ones, and sometimes (as in this case) reintroducing forgotten ones.

### MIGRATION BY JUMPS

Yet even if we can see an alternative worldview in the mindscape that could make our future better, how do we get there? Psychological, economic, political, or organizational barriers likely block most routes from today's worldviews to others that are potentially attractive—just like steep ridges and mountains can block our path across Earth's surface. Our Waterloo group calls this the migration problem.

It might seem relatively easy for a single person to move to a new worldview—say, to switch from a conservative to a liberal political ideology, or vice versa—because, as the systems theorist Donella Meadows said, "all it takes is a click in the mind." But most of us stick with our current worldview, even when abundant evidence shows it's not benefiting us, because it orders our reality and gives our lives meaning. Migration is even harder for a group of people together—say, a community or a whole society—because a group's dominant worldview is always intimately entangled with its prevailing institutions and technologies, and these are usually ferociously defended by powerful vested interests.

Also, migration as a group raises what social scientists call a "coordination problem." Even if everyone involved feels that the current worldview isn't working and has similar ideas about where they'd like to go, those who move first and move alone almost always pay a heavy price. At best, they'll isolate themselves, because they'll be talking in terms most other people can't really understand. At worst, they'll be targeted for attack. Inevitably a lot of people decide to let someone else go first.

Given such obstacles, many scholars, such as the American anthropologist Robert Boyd and biologist Peter Richerson, think people's worldviews and their underlying cultural systems generally evolve incrementally or in small bursts: each slight shift moves us a small step into the adjacent possible, but no further.[10] Then, after our institutions and technologies have shifted a bit too, we can take another small step. By this view, worldview change is necessarily plodding—and large worldview shifts can only happen as small changes accumulate over generations. And because worldview change occurs mainly within single cultures or societies, a global shift, in which much of humanity moves in roughly the same direction more or less simultaneously, is extremely unlikely. This kind of pessimism about fast and far-reaching worldview change encourages many governments and their advisors to look to shifts in institutions and technologies for answers to our global crises.

But our Waterloo group's research suggests that humanity as a whole might be able to jump well beyond the adjacent possible directly to another worldview. In fact, if lots of obstacles block the way between where we are now and that other worldview—a worldview that could be *enough* to genuinely reduce the danger humanity faces—jumping might be the only *feasible* way to get there.

Our research indicates that any big, global jump would require large and simultaneous changes in humanity's answers to several of the fifteen questions in the list, or the addition of entirely new questions that create new dimensions in the mindscape. It would be akin to a Gestalt shift in psychology, or perhaps instances of religious conversion, in which the mind flips from one perspective on the world to another, with no position in between. When we look at the image on this page, our mind jumps from the perception of the female face to that of the male saxophonist and back; our brains find it very hard to hold an intermediate or "in between" state.

At least once before in humanity's history, something like a global worldview jump seems to have happened. Between about 900 and 200 BCE, an era that the German existential philosopher Karl Jaspers labeled "the Axial Age," civilizations in ancient China, India, Israel, and Greece—each suffering enormous upheaval at the time—experienced remarkably similar shifts in their dominant beliefs and values.[11] Scholars still debate the nature of these shifts and why they occurred. Some say they involved changes in moral values: unreflective and absolutist creeds that sanctioned selfishness and violence against members of other communities gave way to worldviews grounded in ideals of self-examination and personal responsibility that emphasized values of compassion, inclusiveness, and fairness. Others say the Axial Age was more cognitive: humans learned how to use abstract knowledge that was disengaged from everyday concerns; people could then see what they believed were more essential aspects of reality—its underlying "truths"—a change that made possible, among other things, modern science and universalized ethics.

Despite their differences on specifics, prominent scholars including the late American sociologist Robert Bellah and the renowned English author on world religions Karen Armstrong agree that the change did occur, that it was of great significance, and that it laid the foundation for modern civilization.[12] Yes, it happened over a period of centuries, perhaps as many as seven. But remember, in those days ideas spread at a snail's pace through societies and civilizations: people traveled, even over long distances, mostly on foot and communicated with each other almost always through speech. Any equivalent worldview transformation this century, in a highly networked world energized by mass travel and modern information technologies, would be immensely accelerated.

Are there other examples in history of global worldview jumps? Two such candidates might be the shift from city, principality, and kingdom to sovereign nation-state as a main unit of human group identity after the Treaty of Westphalia in 1648; or the recent diffusion around the planet of neoliberal economic principles that promote norms of deregulation, economic efficiency, and the commodification of nearly everything. Yet while both these transitions were enormously important, to be sure, neither reformulated the very basis—moral and cognitive—of human civilization, the way the original Axial Age did.

When we view humanity's situation today through a complex-systems lens, it looks like conditions could finally be ripe for a jump of similar magnitude—a kind of second Axial Age. Karen Armstrong herself suggests that such a transition has already begun, starting with the Enlightenment in Europe in the seventeenth and eighteenth centuries. Now, the extraordinary—and historically unprecedented—connectivity, uniformity, feedbacks, and emergence in today's global systems could make possible a deep and rapid transformation in humanity's beliefs about itself and its future—sweeping nonlinear shifts in worldviews that could lay the foundation for a prosperous, just, and even exhilarating new era of human civilization.

### BINDING QUESTIONS

While the state-space model is a good place to start to generate understanding of existing and possible worldviews in the mindscape, it isn't enough by itself, of course. We can use answers to the table's questions to identify some of a political ideology's key assumptions, to see where one ideology sits relative to others, and to find zones in the mindscape that haven't been fully explored. But by themselves, answers to the questions don't give us rich details of the content, or meaning, of a given ideology or its larger worldview. That content is a scaffolding of concepts, beliefs, and values around which people build the stories that guide their lives, including, ideally, the compelling visions of the future that can be the basis for their powerful hope and the hero stories they use to manage some of their deepest anxieties.

To begin to see such details, we must ask a different but related set of questions. For example, if a given worldview entails a belief that some moral principles are objective and universal, we need to ask: Which principles, exactly, does this worldview regard as such? Similarly, if a worldview sees a person's fate as largely a result of circumstances and not choice, we can ask: Which circumstances are most powerful? Perhaps economic conditions during upbringing are thought decisive, or family love, or genetic makeup.

We need answers to such questions to understand the stories people tell using their worldviews. I call the questions "binding questions," because their answers together *bind* a specific location on the mindscape to a detailed picture of what a real or imagined worldview at that location would look like to people holding it in the context of their everyday lives. Binding questions give us an understanding of the worldview "from the inside," so to speak. Below is a sample list.

### EXAMPLES OF BINDING QUESTIONS

Asking questions such as the following of a given worldview or ideology will help establish key details of its perspective. This list is not exhaustive.

**Threat:** If the world is a dangerous place, what makes it dangerous?

**Source of Understanding:** If the world is best understood through reason, what kind of reason (for example, scientific, philosophical, or religious) is most appropriate?

**Spirituality:** If reality is spiritual, what is the locus or source of this spirituality? What is sacred?

**Moral Principles:** If some principles are objective and universal, what are those principles?

**Agency:** If a person's fate is a result of circumstances, which circumstances are most powerful?

**Relationship between Humans and Nature:** If people are exceptional and distinct from nature, in what ways are they exceptional compared to other species?

**Social Differentiation:** If differences between groups are large and essential, what groups and differences are most important?

**Source of Personal Identity:** If one's identity is derived from one's group, which group?

**Time:** If the ideology or worldview is oriented strongly to the past, which aspects of the past are important? If it's oriented strongly to the future, what is its vision of that future?

**Change:** If the ideology or worldview encourages change, what kind of change?

**Care for Others:** If one should help others, whom should one help?

**Authority:** If one should defer to authority, which authority?

**Power:** If the use of power is often right, which entities (for instance, the state, corporations, or paramilitary groups) have the principal right to use this power and when?

**Wealth:** If large differences in wealth are moral, which people and/or groups should rightfully benefit?

Consider the binding question for the issue of Social Differentiation in the table. It's a key one: "If differences between groups are large and essential, what groups and differences are most important?" From their answers to this question, we can learn what the holders of the worldview see as the most important groups in their social environment; they might, for instance, divide up their social world using markers like race or class or religion.

Then we can link together answers to some of the other state-space and binding questions. For example, if a man divides up his social world according to race, we can then ask to what degree race informs his personal identity—especially whether race is the basis for his idea of "we" and, if so, how much that "we" informs his personal identity, his sense of "Who I am." Next we can ask how much he thinks he should care for other members of this "we" group. White nationalists, for example, see themselves as part of the white racial "we" and usually feel strongly that they should take care of other members of what they identify as the white race, to the exclusion of others.

The Social Differentiation question is also important because when we—whatever our orientation—see our social world as cleaved into essentially distinct groups, we often further differentiate among these groups by what we believe is their underlying human nature and degree of agency. Those espousing so-called Nazi ideas (today as much as in the last century), for example, not only differentiate the social world by what they see as "race," they then ascribe to the various races fundamentally different natures, with of course their own appearing superior: Aryans are innately generous, while Jews are innately selfish, according to their grotesque worldview.

One of the most common perspectives on social differentiation simply sees society as divided between groups with lots of power and those with little or none. Those with power—a characteristic sometimes associated with wealth—are seen, often correctly, as having a lot of control over their fate (or agency), which they use to exploit those without power. In the West, people on both the political left and the populist right often see themselves as members of such exploited groups, a view that makes it more likely they'll see the world they inhabit as dangerous. When such beliefs are coupled in their worldviews with strong beliefs in objective principles of justice and fairness—principles that the powerful seem to shamelessly violate—people in both groups can feel tremendous grievance and anger.

We'll see in the last chapter that the questions regarding Social Differentiation, Source of Personal Identity, and the Relationship between Humans and Nature are together particularly critical to shaping humanity's evolving conception of itself—so they're critical, too, to any positive vision of the future that can be the object of our hope. These questions, and all the state-space and binding questions more generally, can help us unpack some of the rich details of alternative visions and their underlying worldviews. But we still need a tool to see exactly how these details fit together and what they mean to each of us emotionally.

And we have just such a tool: cognitive-affective maps.

# 19

# Hot Thought

*We know the truth, not only by reason, but also by the heart.* Blaise Pascal

A COGNITIVE-AFFECTIVE MAP (my colleagues and I call it a CAM) is a kind of concept map—a diagram of the connections between concepts—that helps us see key elements of a person's or group's worldview. The mapping tool was created to reflect recent research in psychology, social psychology, and neurological science that shows emotions play a central role in human perception and decision-making—far greater than previously understood—and are an essential part of almost all the concepts we use when we think and communicate with each other.[1] This work has refuted the long-accepted split between human "cold" (rational) cognition and "hot" (emotional) cognition; it shows, as my colleague Paul Thagard argues, that nearly all our thought is "hot"—imbued with emotion. "Emotions are central to human thinking," Paul writes, "not peripheral annoyances."[2]

A CAM depicts concepts that are closely associated with the person's or group's understanding of a specific subject like nuclear weapons, climate change, or even football. Each concept in the map is assigned an emotional value (or, technically, a "valence") on a scale from highly positive to highly negative. So if we're creating a CAM of someone who loves football, we'll give the concept "football" a highly positive emotional value in their map; if we're creating a CAM of someone who hates nuclear weapons—the way Stephanie May did—we'll give the concept "nuclear weapons" a highly negative emotional value. The concepts are then displayed in a linked network, with each link representing either "concordance" or "discordance" between the two linked concepts, to distinguish between pairs of concepts that, in a sense, support each other and those that are in tension with each other.

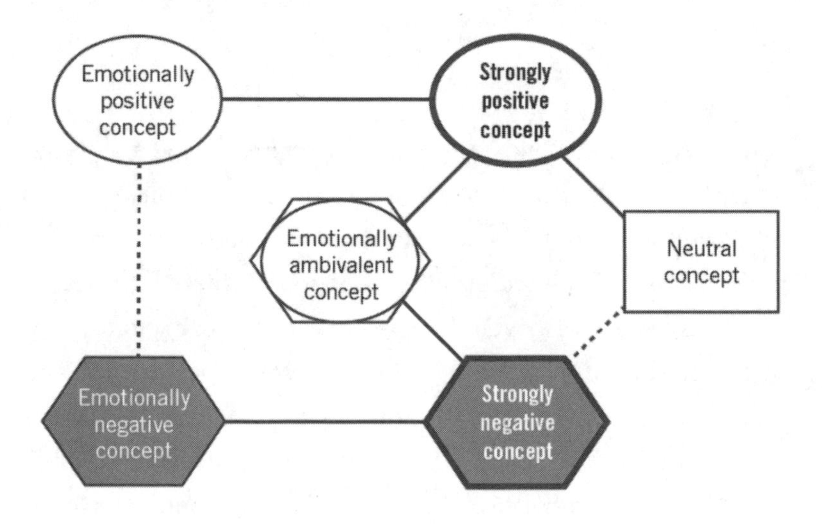


*The basic elements of a cognitive-affective map*

By convention, as shown in the diagram here, we put emotionally positive concepts in ovals (in green, if color is available, but here shown in black and white), and emotionally negative concepts in hexagons (in red, if available, but here shown in gray). The thicker the oval or hexagon's line, the stronger the concept's emotional charge. Emotionally ambivalent concepts are shown by an oval inside a hexagon. Finally, we represent concordant relations between concepts with solid lines and discordant relations with dashed lines. The end result is a detailed image of the core concepts in some aspect of a person's or group's worldview.

The maps are easy for anyone to understand and use. This means that if the people who are being mapped are consulted, they can confirm whether the CAM represents their views accurately and offer corrections if they think it doesn't. My colleagues and I have proposed that people in a conflict can use CAMs to improve their grasp of their own and their opponents' perspectives; neutral mediators can use them too. We've examined diverse disputes, including those over the status of the Western Wall in Jerusalem and housing policy in Germany, to show that CAMs quickly reveal key similarities and differences in disputants' worldviews and point to previously unseen opportunities for compromise or reconciliation; they also make it possible to identify where specific changes in concepts or links might shift the terms of the conflict or create new space for agreement.[3]

And lastly, CAMs make it far easier to see in a glance how the disparate concepts and beliefs in someone's worldview fit together. It's difficult to get this kind of global appreciation with conventional accounts in text, because when we describe someone's worldview with words in sentences and paragraphs, the description unfolds in a linear string of consecutive statements. CAMs, on the other hand, provide an immediate picture—a bird's-eye view, as it were—of a whole chunk of someone's worldview and of the relationships among its parts. (You can find further details on CAMs and an online tool to draw them at www.commandinghope.com.)

## HOW TO USE THE TOOLS

We've now reviewed two tools I believe can help us better understand the inner workings of our own and others' worldviews and better see how worldviews might be changed to support a vision of the future that motivates our hero stories and that's worthy of our hope. The first, the state-space model, gives us a sense for a specific worldview's key assumptions about political and social affairs; the model's binding questions then provide a detailed picture of the key elements—like concepts, beliefs, and values—inside that worldview. And the second tool, the cognitive-affective map, shows us how those elements fit together and their emotional power for the person holding the worldview.

To see how these tools can be used in practice, let's apply them to Stephanie May's story. After carefully reading her memoirs and scrapbooks and talking with her children, the following are my informed guesses about her worldview, using, to start, the state-space and binding questions as my guide.

Beginning with the first question about "threat," did Stephanie regard the world as a safe or dangerous place? It's clear she had very specific fears of nuclear weapons, radiation from testing, and the ever-present possibility of nuclear war. Her apprehension of danger was only occasionally acute, but it remained a constant presence in her mind, penetrating every aspect of her life, and it was one of the main motivators of her activism. Still, she wasn't consumed with feelings that the world immediately around her was dangerous: her writings show she felt secure in her family, community, and broader society. So I think she'd have said the danger was compartmentalized, in a sense, and the world as a whole only "moderately" dangerous.

Going farther down the list, I'd argue that Stephanie's activism sprang, too, from her belief in life's larger spiritual purpose, from her sense of personal agency, from her orientation towards the future and the possibility of positive change, and, most importantly, from a set of basic moral principles that she regarded as objective and universal. The most fundamental of these—her bedrock moral commitment, if you will—was the principle that one shouldn't gratuitously bring harm to children. Although Stephanie wasn't a "strong" moral absolutist—she didn't live by a rigid creed of moral rules—for her, nuclear war and radiation were objective evils, precisely because they'd bring unthinkable harm to children worldwide.[4]

But perhaps the two most significant elements of Stephanie's worldview—both easily overlooked without asking the state-space questions—were her strong commitment to the idea that most people are generous in spirit, or "good" in some broad sense, and her conviction that the similarities between human beings vastly outweigh the differences. Through her long saga of activism, in every encounter with someone new, even with people she had every reason to expect would be fierce antagonists, Stephanie started from the assumption that the person would be reachable through an appeal to their essential humanity. Because she did so, time and time again she brought people over to her side, even some who were at first bitterly opposed to her ideas. And while she could be highly strategic in pursuit of her political objectives, she rarely treated others as instruments to her ends, but instead as ends in themselves.

Stephanie also rejected the idea that there are essential, immutable differences between groups of people. She saw people everywhere as fundamentally the same, so she believed they could all be reached with the same basic moral appeals—especially regarding the need to protect children. Her "we" was the broad public, even all of humanity, and she was powerfully committed to helping all people everywhere, not just those in her own community or nation.

These two commitments—her positive view of human nature and her belief in people's essential similarity—also affected her interpretation of the underlying causes of the nuclear arms race and the bomb testing she hated so much. In her view, these things were happening because of a conjunction of everyday human frailties—fear, incompetence, ego, arrogance, shortsightedness, and psychological inertia—not because of the intrinsically immoral or "evil" nature of particular people or groups. She didn't seek enemies; instead, she strove to understand others and regarded the world's situation as deeply tragic.

With this initial assessment using the state-space and binding questions in mind, we can now draw a CAM to better understand how Stephanie's larger worldview affected her emotional attitudes towards a specific issue like nuclear weapons.

To draw a CAM, it's best to start with a single concept and then identify other concepts that the person you're mapping would immediately associate with it. In this case, I asked myself: What concepts would Stephanie have immediately associated with "nuclear weapons"? The eventual result—after adding, deleting, and moving concepts and links, and referring repeatedly to the materials I had revealing Stephanie's worldview—was the map below. It has two central nodes—"nuclear weapons" (strongly negative) and "America" (positive). Radiating out from each node, almost like spokes on a bicycle wheel, are solid lines to closely associated

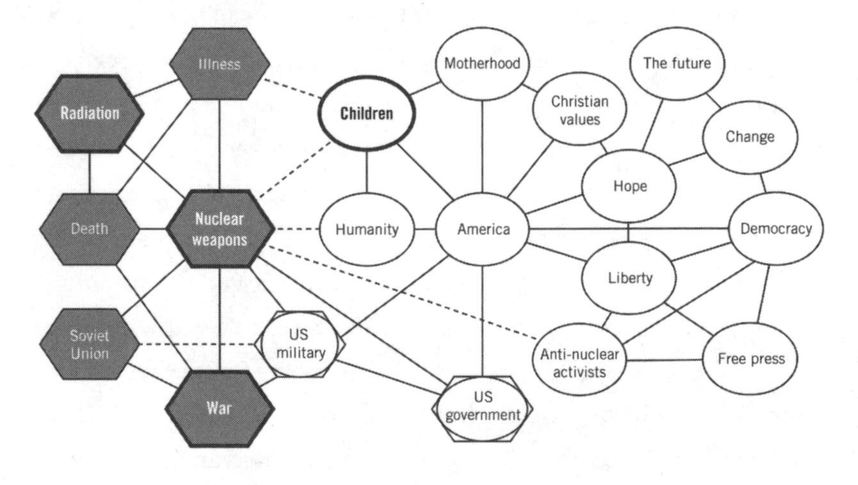


*A cognitive-affective map of Stephanie May's view of nuclear weapons*

concepts—almost all emotionally negative for nuclear weapons, and almost all emotionally positive for America. These links create two large clusters of concepts; and these clusters are then connected, most significantly, by dashed lines between the negative concepts of "nuclear weapons" and "illness" and the positive ones of "humanity" and "children" (the latter the most positive concept of all). The dashed lines represent the huge tension between these concepts for Stephanie, because they were so emotionally opposed in her mind.

It's interesting to compare Stephanie's CAM on nuclear weapons with one we can draw for a representative American anti-communist at the time, such as the patrolman who confronted her outside the Soviet mission, as described in chapter 17.

Once again, we can start with likely answers to the state-space questions and then draw the CAM. Beginning with the first question about "threat," it's reasonable to say that to most American anti-communists in the late fifties and sixties, the world was a profoundly dangerous place, not just because a formidable enemy, the Soviet Union, lurked beyond America's borders, an enemy that they'd been led to believe could incinerate the country at any moment, but also because American society was, in their view, permeated with the enemy's sympathizers. Most anti-communists also believed that America stood largely alone, at that critical moment in history, as the guardian of a set of universal and vitally important moral principles—including the sanctity of life, individual liberty, religious freedom, and the right to own private property and accumulate the fruits of one's labor. They thought these principles were deeply rooted in America's history and culture and not to be challenged, reflecting a worldview bias in favor of the past and against change.

Finally, most anti-communists believed American society could be divided into those who were prepared to defend these principles, and the nation representing them, and those who weren't. Reflecting a strong commitment to individualism and personal responsibility, they believed that competent adults had the agency to choose which side of this epic struggle they were on. American society, by this view, should protect and take care of those people, like members of the armed forces, who chose to be on the right side of this divide, while the government was morally justified in using its coercive power and authority against those, like anti-nuclear activists, who, by this logic, had chosen the wrong side.

This was likely the patrolman's larger worldview context. But in drawing a CAM of his view of nuclear weapons—in this case, US nuclear weapons—we have only a few of his words in Stephanie's memoirs as evidence. So I've interpolated a number of concepts (such as "America" and the "Soviet Union") into his graph for completeness. Since we're postulating what concepts the patrolman would have immediately associated in his mind with "US nuclear weapons," many are different from those in Stephanie's map.

The most obvious contrast between the two maps is the starkly different emotional values for nuclear weapons—strongly negative for Stephanie for all such weapons, but positive for the patrolman in the case of US ones. While the patrolman didn't mention these weapons directly when he spoke to Stephanie, he did say that

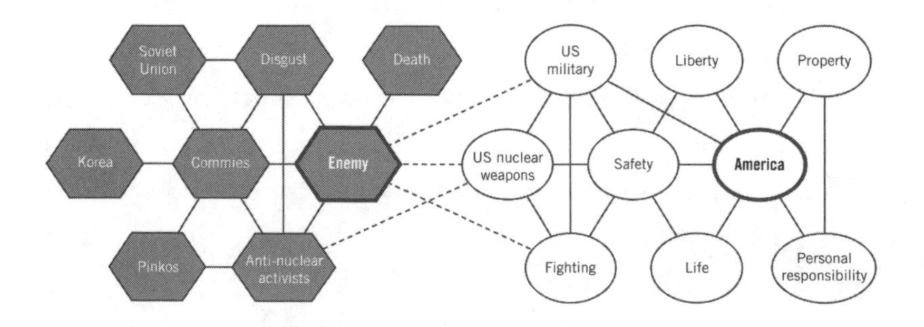


*A cognitive-affective map of the patrolman's view of nuclear weapons*

anti-nuclear activists were "aiding and abetting the enemy." For anti-communists in the early 1960s, US nuclear weapons were keeping America safe; but for Stephanie and her fellow activists, they increased the risk of war, illness, and death.

The patrolman's worldview also personified and unified the sources of danger. Commies, pinkos, and anti-nuclear activists were linked together in his mind, and they were all—one and the same—the "enemy." Nothing was ambivalent in this part of his worldview, which bristles with fear, contempt, and anger. For Stephanie, though, the danger didn't come from a specific group or country at all, but from a specific technology: nuclear weapons. And the negative emotions this technology aroused in her mind were counterbalanced by many sources of positive feeling about her society and possibilities for a better future.

Yet the contrasting CAMs reflect, too, that at that moment in the United States, many anti-nuclear activists and anti-communists agreed on key matters. For instance, many anti-nuclear activists like Stephanie shared positive feelings with their opponents about America as a country and society (even if they were ambivalent about, or even hostile to, specific actions of the US government). Similarly, the patrolman, had he been asked, would surely have agreed with Stephanie's positive associations of America with democracy, liberty, children, and hope.

And while not a matter of agreement (and not fully visible in these CAMs), people in both groups derived vital elements of their hero stories from their worldviews—specifically from ideas about identity, good behavior, fairness, and justice, and from the potent emotions those ideas evoked. In his very first words to Stephanie, the patrolman declared "I was in Korea. I fought the Commies then, and I'm ready to fight them now." He was proud that he was ready to die in the fight against communism, just as Stephanie was proud that she was ready to endure days of ridicule on the sidewalk and possible harm to her health to fight nuclear weapons.

We all need hero stories. But if we're to be motivated, together as a species, to find solutions to the terrifying challenges we collectively face that are both enough and feasible, our hero stories must reflect shared worldview commitments that can serve as a common bond among peoples around the world. These commitments should then be the basis for a shared, compelling vision of a future worth having—a vision that gives us all powerful hope. That vision is my focus in the next chapter.

## 20

# Renewing the Future

*For, in the final analysis, our most basic common link is that we all inhabit this small planet. We all breathe the same air. We all cherish our children's future. And we are all mortal.* John F. Kennedy

KATE STOPPED RUNNING AND, with tears streaming down her face, turned to me and cried: "Will this story have a happy ending?"

We were on a hike in the woods, and just a few minutes before, Sarah had become separated from the rest of the family. Fearing her mother was lost, Kate had bolted down the trail, with me running behind, trying to reassure her.

The story did have a happy ending, of course. Mother and daughter were soon joyously reunited. But I'll never forget the question Kate asked, or the expression on her six-year-old face when she asked it.

Adults learn not to ask this question, of course, at least not so explicitly. We know that true happy endings are terribly rare. We know, too, that things never really "end," though we do. One process just unfolds into another in a continuous, infinite stream. Still, we can't escape the question entirely, because all human beings—not just children, but adults too—see and think about the reality around them in terms of stories. Often these are hero stories—about ourselves or about our families, communities, and nations—with protagonists (good and evil) acting out plots across rich emotional landscapes. Our stories use the conceptual scaffolding of our worldviews—especially the beliefs, values, and goals we derive from the groups we identify with—to convert the meaningless material and social reality around us into a meaningful and even spiritual world of intentions, reasons, and, most important of all, purposes. For adults, a happy story is one that offers a clear and appealing purpose for our lives and gives us a sense of agency to realize that purpose.

But now, around the world, fear is causing many of us to pit our stories against each other, eroding humanity's already diminishing sense of collective purpose and feelings of common identity.

So, what story about the future should we tell our kids—children who will live as adults in an age of loss that will accentuate that fear? What positive vision of the future can we create for them, and for ourselves?

If I'm honest, I can't tell Ben and Kate a story that has anything like a happy ending. I can, though, tell them a story about their stories, by providing them with some ways to make their stories—and ours—more meaningful and helpful in their lives, despite the future's likely loss and fear. That's what I've tried to do in this book, which is, in a sense, my "how about" for my children.

In my conversations with them in coming years, we'll use our recursive imaginations to travel together to a difficult future, but one in which they're nonetheless living purposeful, meaningful lives. In that future, they'll have related their own, personal hero stories to the monumental challenges that humanity faces, and they'll have empowered themselves to engage with those challenges, in part through their honest, astute, and powerful hope. They will be commanding hope.

In a way, then, their stories can be happy. They can certainly be heroic, because all our children will live and participate in one of the most turbulent, exciting, and decisive periods in our species' history. Everything will be on the line, and nothing will really matter except working together to find a feasible path to a shared vision of a positive future. And just like their parents now, all our children should be able to make a real difference in whether that vision is realized, if they believe they can.

For Sarah and me, the job of showing our children that they can make a difference has already begun. As they've learned more about what's happening in the world, they've wanted—they've urgently needed—to do something in response. Ben has started organizing his classmates to protest inaction on climate change—as part of the global School Strike movement—while Kate has joined an online community to protect biodiversity (specifically, the piping plover, a charming but threatened North American shorebird). These concrete activities are helping them cope.

I think we can help our children, too, through the examples we set. Stephanie May's children, Elizabeth and Geoffrey, watched her anti-testing activism and sometimes joined her. In the photo on page 15, six-year-old Elizabeth is standing beside her mother as she addresses the crowd in London's Trafalgar Square in 1960. Decades later, Stephanie wrote about why it's so important to help our children believe they can make a difference:

> Although I didn't think about it at the time, I believe in retrospect that encouraging Elizabeth and Geoffrey to become a part of the work closest to our hearts not only made them feel more secure about the world but also gave them a belief that the individual is not helpless to change things, to right wrongs, and to affect government policy. Today, both Elizabeth and

> Geoffrey are in the forefront of the environmental struggle to save the earth.[1]

Alas, in 2020, one story about our collective future that's becoming increasingly ubiquitous—and maybe even appealing to those inclined to resignation—starts with the opening line "we're doomed" (or, in the vernacular, that we're "fucked" or "screwed"). I hear this kind of declaration from a substantial and rapidly growing proportion of the students I teach. I respond to these young people by saying, as I say to Ben and Kate, that humanity is in fact doomed only if we collectively *choose* to be doomed.[2] But I also caution that if we're to choose otherwise—especially if we're to choose to move towards a genuinely positive future—large numbers of us around the world will have to invest enormous effort and intelligence, and show extraordinary courage, to make that better alternative real.

Opportunities for such choice and action are appearing already and many more will appear soon: the pulses of social earthquakes that are going to increasingly affect our world will create openings—if we're strategically smart—to shift our collective path in more positive directions. Young people around the world are seizing these opportunities. Real social and political change only happens in times of crisis, because crisis is needed to discredit existing systems of worldviews, institutions, and technologies, and the structures of power that sustain them. This weakening (or even breakdown) of the current order creates possibilities for "catagenesis," as I called the phenomenon in my last book, *The Upside of Down*, which is a process of rebirth through breakdown.

Such times of crisis supposedly bring out the best and worst in people. But I think it's more accurate to say that crisis tends to sort people by moral character—the kindhearted person from the cruel, the honest from the dishonest, and the brave person from the cowardly, for example. This century will see a great deal of such sorting. It's just as important to remember, though, that crisis isn't antithetical to hope; indeed, it can galvanize hope and the fortitude that justifies that hope. In the dark days of late May 1940, as Hitler's forces closed in on British and Allied troops at Dunkirk, and England faced the real prospect of Nazi occupation, the British writer and theologian C.S. Lewis wrote to a friend: "Oddly enough, I notice that since things got really bad, everyone I meet is less dismayed."[3]

Fundamentally, the story we tell our children and ourselves must be about what we should do—today, and as the future unfolds. The social and political changes needed will only come about through aggressive activism, of countless types and forms, but all intensely political. Mobilizing large numbers of people across diverse societies around a positive vision of the future is a political project—and will likely be an increasingly dangerous one, too.

Does Stephanie May's story tell us anything relevant to guide this present and future activism? One could contend that her activist model is woefully outdated—that, for example, her faith in people's basic goodness is dangerously naive now. But I believe, on the contrary, that Stephanie stands as one example of a vital archetype of the kind of activist we need, and that Greta Thunberg and her brave fellow climate protesters around the world are Stephanie's direct descendants in terms of their forthright honesty, strategic astuteness, and awareness of the impact of the dramatic political gesture.

They have hope themselves, and through their actions they offer all the rest of us hope too. But for that hope to be truly powerful, it must incorporate a compelling vision of the future. So in this last chapter, I outline my own idea of such a vision—of a world to come—that we have every chance of making possible tomorrow.

**VALUES AND TEMPERAMENTS**

A compelling vision is one that appeals to our shared values as human beings and, also, to our common personality temperaments.

I see human values as coming in three main types: utilitarian, moral, and existential. What I call "utilitarian values" reflect our uncomplicated likes or dislikes—for vanilla ice cream over chocolate, for instance. Most economists use this notion of value; in their parlance, our "preference" for one thing over another thing is determined by its greater "utility" to us.

But this notion of value reflects an astonishingly impoverished understanding of human beings: it's based on the assumption that we're nothing more than decision-making machines choosing among options based on how much they satisfy our hedonistic desires. It underpins economists' common assumption that we're insatiable consumers of material stuff and that this consumption is key to the good life. To maintain economic growth, mammoth industries of persuasion have arisen to foster mass consumption regardless of its actual importance for our well-being, and a huge apparatus of economic theory has made this consumption morally legitimate purportedly to sustain the "health" of the economy. According to this view, people are little more than walking appetites. Thus the Consumer Society.

Yet we're obviously so much more! We also have "moral values," which are emotionally charged rules or principles that we believe ought to govern our conduct, most often with other people, but also with other living creatures. These "oughts" can be specific injunctions, like "we shouldn't hit each other," or more general principles like the Golden Rule. Many people believe such "oughts" ultimately come from a higher authority, such as a deity in which they have faith. They consider these injunctions as objective, universal, and absolute, and hold, as part of their belief, that they'll be punished if they violate them. Others think their moral values arise from their community's history and culture, or that they simply make sense given the evidence they draw from human interactions. They generally don't believe these values are objective and universal, and don't have a clear idea of how or even whether they'll be punished if they violate them.[4]

Finally, we have what I would call "existential values" (although some might call them "spiritual values"), which help us answer the big questions about why we're here and what our purpose is. These values concern, most deeply, how we understand our specific role in our larger story about reality—about our relationship to the cosmos, as it were. We typically don't spend a lot of time talking about these values, and many of us turn to religion for them; but in some ways they're the most important values of all, because they're key to our views about what makes our lives meaningful and good.

Very roughly, as we move from the bottom to the top of Abraham Maslow's hierarchy of human needs (chapter 9), we start with utilitarian values at the base, shift to moral values in the middle, and ultimately come to existential values at the top. We derive our moral and existential values mainly from the worldviews of the groups that matter to us, and we use them to create our vision of a desirable future and the hero stories and immortality projects we construct for ourselves within that vision. They are fundamental to the meaning of our hope.

I also see three general types of temperament in the human population: exuberant, prudent, and empathetic. I've developed this three-part distinction—shown in the table on the next page—through my work on ideologies and worldviews, including the state-space model, and especially after reflecting on contentious public debates over environmental issues. It's at best a rule of thumb, and it's not wholly original; but I still find it extraordinarily useful.[5]

By temperament I mean a person's underlying emotional and cognitive predisposition towards the world. People with an exuberant temperament are fundamentally optimistic and happy. They have a strong sense of human agency, so they're joyous about life's possibilities to explore, create, and flourish. And they aspire to have the opportunity to change and to grow, which means they're deeply averse to any kind of constraint.

People with a prudent temperament, in contrast, are fundamentally cautious. They have an acute sense of the dangers lurking in the world, so their main aspiration is safety. They tend to be more skeptical about human agency and more motivated by fear, and they're apt to think recklessness and profligacy tempt fate. But their perspective isn't necessarily bleak: because they're sensitive to the fragility of order and to the interdependence of things around them, the world often provokes in these people feelings of awe and reverence.

Finally, those with an empathetic temperament are motivated by both compassion for people (or perhaps for living things generally) and by anger when the people (or living things) they care about aren't treated well. They're averse to suffering and, most fundamentally, they aspire to justice and fairness as understood within a framework of strong moral values.

| Temperament | Dominant positive emotion | Principal aspiration | Aversion to |
|---|---|---|---|
| **Exuberant** | Joy | Opportunity | Constraint, limits, pessimism |
| **Prudent** | Awe | Safety | Danger, recklessness, profligacy |
| **Empathetic** | Compassion | Justice, fairness | Injustice, unfairness, suffering |

*Three common temperaments and their key properties*

Several of the state-space questions I introduced on page 318 highlight key differences between the three temperaments. People with an exuberant temperament will tend to see the world as safe, believe in agency, be oriented towards the future, and be inclined to encourage change and resist authority. People with a prudent temperament are more likely to see the world as dangerous and be skeptical about agency. And finally, people with an empathetic temperament will stress the basic generosity of human nature, the fundamental similarity of all human beings, and the importance of caring for others.

Most people's personalities mix all three temperaments to different degrees. If an apex of a triangle represents the pure form of each, hardly anyone falls exactly at one apex; instead people generally fall somewhere in the space in between the three. And to the extent that most of us have a bit of each temperament available to us, we're able to emphasize one bit or another, as our circumstances change. But what has surprised me over the years is how readily people can be sorted into one of these three categories—how much, in other words, people tend to cluster towards the triangle's corners.

That's partly because the temperaments sometimes don't combine easily. It's especially hard to mix the exuberant and prudent temperaments. Exuberant striving for opportunity and self-expression doesn't easily fit with prudent awareness of danger and constraint, so these two temperaments can be like oil and water. It's much easier to mix the exuberant and the empathetic or the prudent and the empathetic.

This tripartite distinction isn't simply another way of labeling the standard left-right ideological spectrum. Sure, people who are exemplars of exuberance, such as capitalist entrepreneurs, are often conservative, but they are neighbors on the ideological right with more prudent conservatives who emphasize restraint, stability, and caution. Similarly, while many empathetic lefties strive for their version of justice, some are exuberant activists while others are cautious and careful.

And of course, none of the three temperaments is good or bad in any absolute sense. As human beings, we appear to have evolved these three psychological tendencies because each, depending on context, serves a vital social purpose. Sometimes our societies need to be warned of danger; other times, we need all the exuberant agency we can marshal to innovate or respond to a crisis; and at yet other times we need to be reminded that our members' well-being is paramount.

This categorization of temperaments is simple but, among other benefits, it helps us understand people's reactions to the issue of economic growth and its possible limits. Most people who are emotionally invested in economic growth—and I'd include in this group most members of Western corporate elites and many techno-optimists—have exuberant temperaments, and many of them have built their hero stories around the freedom for agency and personal flourishing that growth provides. Their reaction to the idea of limits to growth often resembles their reaction to the prospect of heavy state regulation of free markets. Both represent a kind of death—in this case, the death of the spirit of agency and opportunity. (Thus the common anti-regulation injunction: "Remove the dead hand of the state from the market!") Many people who are critical of economic growth, on the other hand, are environmentalists with prudent temperaments; they see growth as recklessly damaging Earth's natural systems, and for them, it's growth, not the absence of it, that represents death.

These simple categorizations of temperaments and values can also help us make our vision of the future—and the hero stories we weave within that vision—more powerfully motivating for a broad swath of humanity. Today, our vision and stories must appeal to much more than our hedonistic, utilitarian values, for instance. They need to articulate clear moral and existential principles that position us all in a larger narrative of social purpose, while giving us guidance for what's right and fair.

Yet today's dominant economic worldview, which sustains a global monoculture of consumerism in all Western societies—and inspires the development of a like monoculture in new middle classes in non-Western societies as diverse as Brazil, South Africa, India, and China—appeals mainly to utilitarian values and treats us as atomized, purposeless individuals. Its vision of the good life is represented in countless advertising images of personal physical pleasure: lounging on a tropical beach, living in a luxury mansion, vacationing on a cruise liner, driving a fast car along a winding mountain road, and so on. The images signal the high social status of the people enjoying these pleasures; if you've "arrived," if you're rich enough to be living this way, the images tell us, you must be at the top of the social hierarchy.

Not only is this economic worldview now radically at odds with Earth's deteriorating material reality, but it also doesn't remotely meet people's full range of psychological needs. Once our basic physical needs at the bottom of Maslow's hierarchy are met, consumption of further material stuff is far down on just about everyone's good-life list. Give us happy loved ones, supportive community, a satisfying group identity, rewarding work, moral purpose, some control over our destiny, and reasons for hope, and you'll give most of us nine-tenths of what we really desire in a good life; at that point, looking down on everyone else from the top of the material-consumption status hierarchy becomes much less psychologically satisfying and important.

Still, facing a future that looks treacherous, and without a worldview or a vision of the future built around clear moral and existential values that give our lives meaning, it's easy to understand why we might try to assuage an amplified death anxiety through further consumption. There's nothing to beat "shopping therapy" at certain low moments. Research shows that in a world that seems out of control, rather than hunkering down and reducing consumption, people often just buy more stuff, because it's something they think they can control; and, at the very least, it gives them an immediate burst of utilitarian happiness. The irony is obvious: the worse that problems like climate change become and the more scared we get, the more stuff we'll consume . . . another vicious circle.

### DIVERSITY IN THE BASIN

In sum, for our vision of a humane future to be compelling—and for this vision to give us all grounds for truly powerful hope—it must appeal broadly to our full range of values and temperaments; it must also link these values and temperaments to superordinate goals for all humanity that reflect our world's harsh new material and social realities; and, finally, it must enable people and groups the world over to define for themselves motivating hero stories in which they work together towards solutions that are both feasible and enough to reach those superordinate goals.

In our search for such a vision, we'll benefit if we have a clear idea of what's likely to be its main counterpoint—the most powerful alternative but *inhumane* vision. For our hope to be strategically astute, we need to keep this potential opposing view clearly in mind, grasp its internal meanings, and understand why it will likely have a strong emotional grip on its adherents.

This counterpoint vision will probably be anchored in the kind of "Mad Max" outlook I described in chapter 3. As worsening stress, scarcity, and loss create conditions ripe for perspectives that eschew reason and promote intolerance and violence, a cluster of Mad Max worldviews will coalesce into a deep basin in the mindscape. If we use the state-space questions to tease out some key assumptions, as I demonstrated in the last chapter, we find that these worldviews will probably depict the world as an extremely dangerous place, where moral principles are contextual and determined by who's got power (might is right), fate is a result of one's circumstances, human nature is wholly selfish, the use of power over others is usually right and, most importantly, differences between groups of people—defined by factors like race, ethnicity, or religion—are large, essential, and immutable. In contrast to the hyper-individualism of the protagonist's worldview in the *Mad Max* movies, however, I think these future worldviews will be somewhat communitarian. Dangerous circumstances usually cause people to pull together into strong identity groups. In this case, the group identity in question will likely resemble far more the blood bond of a criminal gang than the civic identity of, say, a nation. And combined, these beliefs and values will give the adherents of such worldviews the powerful conviction that they're licensed to use ruthless force to protect themselves and their group from any conceivable threat.[6]

As we search for a humane alternative to this hideous outlook, our first instinct might be to take a worldview "off the shelf," rather than to create something that's new or even partly new. The most appealing choice might seem to be one of today's conventional left-of-center political ideologies. These perspectives—which in the United States, Canada, and Europe have emerged as among the most vigorous counterpoints to rising right-wing authoritarian populism—generally see social phenomena through the lens of economic class and injustice and regard unbridled capitalism as the cause of nearly everything that ails our world, from economic inequality to Earth's environmental crisis.

Many of these ideologies have their roots in economic and social circumstances—the industrial and manufacturing economies of the nineteenth and early twentieth centuries—when it made much sense to see society as cleaved into capitalist overlords and toiling workers, and to see these two classes as locked in inescapable antagonism. Today, in some quarters the idea persists in an almost Manichean distinction between capitalists as evil and just about everyone else as noble. And while today's anti-capitalist worldviews do usually recognize the gravity of the environmental crisis, they also generally see economic growth as highly desirable, because growth is good for workers and the poor. And for people with such views, when growth and jobs clash with environmental protection, in practice the interests of workers and the poor usually trump those of nature.

The claim that capitalism is the cause of our environmental problems is only partially true, at best. Historically, non-capitalist economies, like that of the Soviet Union, have also caused massive environmental damage; and environmental problems like climate change always have multiple causes—such as people's psychological tendency to discount future costs—many of which have nothing to do with capitalism.

These facts suggest that anti-capitalist worldviews and the visions of the future they support aren't our best starting point to respond to our new material reality of rapid, global deterioration in natural systems and steeply rising risk of environmental catastrophe. Their divisiveness also seems to fit badly with our desperate need to pull together around the world to solve our common problems. Anti-capitalist worldviews and visions may identify someone to blame for the mess we're in, and they might allow us to build our hero stories around the never-quite-realized prospect of a revolution to dismantle the capitalist system, but people holding them should be careful what they wish for. In any direct, knock-down-drag-out fight in coming decades—after the interlinked set of worldviews, institutions, and technologies that justifies and motivates capitalist growth is discredited and falters—those who hold versions of the Mad Max worldview will have one big advantage over the rest of us: they'll feel no moral compunction against using utterly ruthless violence to win.

In our search for alternative visions of the future and worldviews that can guide us and offer us hope, we can rightly be cautious about recycling versions of past ideas. The challenge is to be open to substantially new ones. It's also important to recognize that humanity doesn't need anything like a universal consensus around a single, precisely defined alternative worldview or even a single alternative vision of that future. Such consensus is neither possible nor desirable. Despite forces encouraging the global homogenization of cultures in recent decades, we're still far too diverse to agree on anything like a single worldview. And thank heaven, because diversity is a source of resilience. Diversity in human culture, anchored in our local, regional, and national ways of living, enlarges the reservoir of knowledge, values, resources, and strategies that we can combine in innovative ways to respond to the rapid change and surprises ahead of us all.

At the same time, there's a clear and present danger that too much diversity, or the wrong kind of diversity, will only produce weakness through fragmentation. In confronting the brutal Mad Max worldviews clustered together in their deep basin in the mindscape, our positive alternatives need to be centered on a *core set of generally shared principles*—principles that recognizably link our common temperaments, moral intuitions, and values to our species' superordinate goals and a rough outline of a shared vision of our desired future—so we create together our own broad and deep basin.

Four interlinked principles are key, I believe: they concern opportunity, safety, justice, and identity. This is a moment to come together around these principles, to reinforce the social solidarity we need to thrive together on a crowded, grievously damaged planet.

### OPPORTUNITY, SAFETY, AND JUSTICE

The renowned economist Amartya Sen has proposed a "Capability Approach" to human well-being.[7] By this view, "capability" is the real-world freedom a person has to choose life outcomes (what Sen calls "functionings") that he or she values. The first three of my principles—opportunity, safety, justice—map closely onto Sen's approach: the principle of opportunity corresponds with Sen's notion of capability, while the principles of safety and justice concern the range of outcomes (functionings) available to a person.

The principles of opportunity, safety, and justice also map directly onto the three temperaments I've identified: opportunity onto the exuberant temperament, safety onto the prudent, and justice onto the empathetic. If people across all human cultures generally sort themselves into these three temperaments, then our visions of the future (and associated worldviews) must appeal directly to all of them, if they're going to help bring people together.

Some might think that a core commitment to *opportunity* makes sense only within a Western individualistic culture and that, in fact, even talking about this commitment is just Western navel gazing. But I've found in decades of observations across many cultures that the idea of personal agency is truly universal. It's understood in subtly different ways in different cultures—individualistic freedom being only one—and it's often strongly circumscribed by other commitments to, for example, what is considered honorable in a given culture or to the well-being of the broader community. But personal agency is still recognized everywhere as a fundamental desirable characteristic of people's lives. It's inescapably real to every one of us, after all, in our conscious awareness of our ability to make choices. And everyone—regardless of their culture or circumstances—wants to expand the scope for expression of his or her own creativity and agency as much as possible. That's in our human nature.[8]

If we want to encourage folks with exuberant temperaments—including open-minded capitalists—to create hero stories in which they see themselves inventing, exploring, and experimenting with ways to solve our world's critical problems, then our alternative worldviews and our vision of the future must provide opportunity for personal agency to flourish.

A core commitment to *safety* is essential too—the physical safety of ourselves and our loved ones from bodily violence and other dangers, but also mental safety from psychological trauma—and not just to bring on board those of prudent temperament. When we don't feel safe, we become afraid; and when we're afraid, we often become less trustful of others and less willing to cooperate with them, which makes it hard if not impossible to sustain broad social commitments to the principles of opportunity and justice.

In extreme situations of insecurity and fear, we dehumanize other groups—we no longer see the people in them as individuals, each with distinct and complex characteristics, histories, and goals. And by denying the legitimacy of another group's ways of life, interests, actions, and even existence, we also exclude its members from our shared moral community, and so from the protection afforded by our community's moral constraints.[9] This kind of psychological flip makes it easier for people to fight and imprison or kill others seen as threats, and it's a key feature of the most pitiless acts of human violence. It's also exactly the emotional dynamic that drives the Mad Max worldview.

In many areas of the world, social earthquakes are already frightening people enough to turn to authoritarian leaders who promise protection through the hard, uncompromising application of state power. Authoritarianism undermines key institutional elements of safe societies—especially the rule of law and uncorrupt institutions—while also compromising our opportunity and personal agency. A commitment to safety is essential if we're to stop this process—and preserve space for some form of democracy, broadly defined as the values, institutions, and practices that ensure a voice and role for every competent person in the governance of their society.

The challenge of sustaining and reinventing political democracy in our hyper-stressed world is, along with the challenge of reinventing our economies in a post-growth form, among the most perplexing we face. Our current democratic institutions evolved mainly in the nineteenth and twentieth centuries, when most information relating to public affairs was channeled through print or broadcast media, most individuals had very limited abilities to communicate with large numbers of other people, and the problems our societies faced generally developed far more slowly and in less complex combinations than they do now. In today's radically different and rapidly changing circumstances, almost all democratic institutions around the world are under stress, and no one really knows how or even if they'll adapt.

Because social earthquakes will worsen dramatically in coming decades, any viable, humane worldview must meet the challenge of rising authoritarianism, and the violence it often expresses, head on. While our alternative worldview shouldn't encourage us to initiate violence, it should give us a framework of beliefs, values, and emotional responses that prepares us to respond to violence when and if necessary—effectively but within moral restraints. It should license us to defend ourselves and our communities and societies, while incorporating beliefs and values to support the institutions—in particular an inclusive rule of law—that guard against dehumanizing people outside the "we" group and that preserve space for opportunity and democracy.

A viable worldview that helps sustain an equitable society should have a core commitment to *justice* as well, to reach people with empathetic temperaments, and also to support our commitment to safety. We all feel humiliated and angry when we think we've been unjustly treated, and these feelings, when shared by many in a society, weaken social stability. A commitment to justice directly supports our commitment to opportunity, too, because it requires us to address the often repressive effects of social power in our societies—from the overt activities of vested interests to the more subtle influences of language and political and economic institutions—that block positive change and limit people's chances to flourish.

We all carry some idea of justice in our minds, just as we all have moral values. For some, justice is achieved when people follow certain moral rules; it's a matter of the right kind of social order and behavior and of who deserves rewards and punishments within that order. For others, justice is achieved when the distribution of opportunities, power, and wealth among people in society is fair—according to a general principle of, or intuition for, fairness. I see justice as encompassing what ethicists call "deontological" principles—or what most of us think of as moral "bright lines"—that establish people's fundamental moral responsibilities and rights as human beings. One of these principles is the injunction, proposed by the philosopher Immanuel Kant, that people should be treated as ends in themselves, not as instruments to one's ends. I see justice as encompassing, too, basic principles of fairness regarding the distribution of economic and social benefits and opportunities (wealth, status, and the like) within and across our societies.[10]

While everyone gives lip service to the general idea of justice, we tend to disagree deeply about the status of justice-related principles—whether, for instance, they're universal and objective—and about what counts as a just social arrangement. I've tried to address these challenges earlier (chapter 13), partly by proposing three bottom-line injunctions that nearly everyone should be able to accept:

don't wreck our planetary home, don't commit mass suicide by fighting among ourselves, and protect our children. I regard these three injunctions as profound moral obligations—bright lines—for humanity collectively and for each of us as individuals. (Stephanie May saw the third, of course, as the moral basis for her anti-testing activism; and all three injunctions are at the heart of today's climate marches and strikes.) They're so obviously valid that we can take them as universal and objective (even without any reference to a higher authority, such as a deity).

Over the last two centuries, humanity has arrived at a number of similar moral commitments. For instance, we've come to see genocide, slavery, child prostitution, and chemical warfare as unequivocal wrongs. These moral commitments have been codified in international treaties and laws, such as the 1948 Convention on the Prevention and Punishment of the Crime of Genocide. State and non-state actors often violate these treaties and laws, of course, and many scholars now fear that the entire institutional edifice which they created and through which they're enforced—from the International Court of Justice to the International Labour Organization—is weakening fast. But the underlying moral principles on which this edifice resides retain broad and deep support across the human population.

For some bizarre reason, though, our species hasn't yet come to see an action like destroying our planetary home in the same way. Nonetheless, that outcome would be an unparalleled moral calamity, an unmitigated evil, and not just because of its consequences for us and our children, although those are reason enough. It would be a moral calamity, too, because Earth is also home to an uncountable number of other living things, and we have no plausible moral right to drive masses of them to extinction.

In short, what we're doing collectively to destroy our planetary home is just wrong, plain and simple. Pope Francis used his second encyclical *Laudato Si'* in 2015—titled "On Care for Our Common Home"—to forcefully articulate the moral basis for such a deontological principle, which also underlies the Earth Charter, drafted by Maurice Strong and Mikhail Gorbachev and released in 2000.[11] But these vital statements haven't yet elicited broad consensus across humanity, even so late in our planet's unprecedented environmental crisis.

Coming to a consensus on any notion of justice as fairness is going to be even harder. But we can't escape matters of fairness. Global problems like climate change are ultimately problems of planetary equity—and of our world's enormous gaps in wealth, power, and (again) opportunity—because the rich and powerful have usually played a far bigger role in creating these problems, while the poor and powerless are usually first to suffer the consequences.[12] Agreeing on any notion of justice as fairness is made even more difficult if one considers historical inequities. Does "justice" require, for instance, reparations for slavery or for Indigenous peoples' loss of land to settlers around the world?

Very slowly, far too slowly, humanity is grappling with such fairness conundrums. In the international law emerging around climate change, for example, countries have "common and differentiated responsibilities" to address the problem: all countries have a responsibility to do something, but a country's degree of responsibility varies according to its contributions to climate change and its capacity to respond. That means countries that have become wealthy by burning gargantuan amounts of fossil fuels, like the United States and Canada, are obligated to do far more to fix the problem than poorer countries.

## OUR COMMON FATE

The fourth principle that should be at the center of any positive vision of the future and worldview is a strong commitment to a shared identity that encompasses not just all of humanity, but in some respects all life on the planet too. This is the matter of who we see as "we." I've raised it repeatedly in this book, because it's so critically important. In some ways it's the glue that holds everything else together.

In my studies and travels over the years I've noticed that thriving societies generally share one characteristic: their members—high and low, rich and poor, powerful and weak—have a strong sense of community that encompasses most if not all of that society's members. This identification with and commitment to the commonweal is rooted in people's perceptions of a common fate—the emotionally charged belief in a shared destiny. It provides the basis for the development of what social scientists call "social capital," which consists of strong bonds between people of mutual trust and reciprocal obligation (where, in its most basic form, a person feels obliged to help someone because that person has helped them previously). Research shows that communities and societies with lots of social capital are better at solving their problems big and small.[13] There's no obvious limit to the number of people who can be included in the ambit of a "we" identity; some of the world's largest "imagined communities"—India, China, and the United States, for instance—encompass hundreds of millions of people.[14]

All of which has led me to the conclusion I stated in this book's first pages: humanity can't and won't address its urgent problems, like climate change, until enough of us from a broad range of cultures think of ourselves as facing these problems together, as members of a common "we."

A pious Western liberal platitude? Perhaps, but it also reflects the hard-headed realism required of the moment. The importance of a common "we" to human cooperation is one of those basic realities of our social world that profoundly shapes our possibilities. Because our critical problems are now global in both their causes and scope, we need broad collective action across the world's societies to address them successfully. But as the leaky-lifeboat metaphor shows, until enough of us think of ourselves as part of one community on this planet, we won't be able to make the difficult choices about who's going to do what to help whom, or about when and how wealth and know-how should be moved from one part of the planet to another. A fractured humanity—seeing everything through the Mad Max lens and split into distrusting ethnic, religious, and cultural tribes—can't possibly solve global problems and will ultimately fall, divided.

The world's disjointed political systems, particularly its nation-state system, are at the heart of our cooperation challenge. Any country that takes the first major steps to address a problem like climate change—by leveling heavy taxes on carbon emissions, for example—risks bearing huge costs and putting itself at a competitive disadvantage, while generating little immediate benefit for itself and others.[15] Powerful vested interests in each country—fossil-fuel interests like oil companies, in the case of climate change—use this collective action problem as an excuse to declare it's in their nation's economic and security interest to delay action. They exploit, too, people's legitimate fear of limits on their nation's economic growth and their distrust of "big government" regulation and meddling (or, more ominously and less legitimately, fear of "world government" domination—a fear that's especially prevalent among right-wing groups in the United States).

Social scientists have studied how to overcome these obstacles to broad cooperation.[16] They've shown that in such situations, given enough time, narrowly self-interested actors, like nation-states, can learn to cooperate to achieve outcomes that benefit everyone and to build the institutions—the accepted rules of the road—needed to support cooperation going forward. The international trading regime is perhaps the best example of this kind of outcome. But in 2020, humanity doesn't have the luxury of going through a painful, half-century or longer process of learning and institution-building. We face catastrophe in at most a few decades and perhaps much sooner. Our boat is about to sink.

But we also know from our scientific research that having an encompassing "we" identity—one that extends across all actors in the system—makes cooperation vastly easier. It can literally cause collective action problems to dissolve into thin air. The social and cultural psychologist Jonathan Haidt provides a wonderful example when he describes how rice farmers in Bali, who are divided into small family cooperatives called "subaks," dealt with the challenge of distributing irrigation water among themselves.

> The ingenious religious solution . . . was to place a small temple at every fork in the irrigation system. The god in each such temple united all the subaks that were downstream from it into a community that worshipped that god, thereby helping the subaks to resolve their disputes more amicably. This arrangement minimized the cheating and deception that would otherwise flourish in a zero-sum division of water. The system made it possible for thousands of farmers, spread over hundreds of square kilometers, to cooperate without the need for central government, inspectors and courts.[17]

Here, the "we" identity that the subaks established in their minds as they worshipped a common upstream god made it possible for them to cooperate around water management. But what might be the psychological analogue of an upstream temple god for humanity today? I think there are two possibilities.

The first is the reality that all parents care about their kids and want more or less the same things for their kids. This truth holds across all divisions of race, ethnicity, religion, language, caste, and class, and even across the civilizational divisions—between West and East, or between Islam and Christendom—that some fear are now insuperable. People want a world—and a future—in which their children are safe and secure and in which they can flourish as human beings. They want, in other words, circumstances for their children that are governed by general principles of opportunity, safety, and justice. People use different language for these principles depending on their culture—and the details of what it means to "flourish" obviously differ, sometimes a lot—but the underlying sentiments are universal.

If we can agree on something so fundamental across such different cultures, groups, societies, and civilizations, maybe—just maybe—we have a kernel of an idea around which we can develop the feeling of global identity we need to address our shared problems. "Protect our children" must be one of our bottom-line injunctions. Indeed, the young activists engaged in strikes for climate action around the world are angrily demanding that adults and their parents respect precisely this injunction—that they protect the children, them. As Greta Thunberg passionately declared when she—along with fourteen-year-old Alexandria Villaseñor and fourteen other young activists aged between eight and seventeen representing Argentina, Brazil, France, Germany, India, the Marshall Islands, Nigeria, Palau, South Africa, Sweden, Tunisia, and the United

States—formally filed a complaint on September 23, 2019, with the United Nations Committee on the Rights of the Child: "World leaders have failed to keep what they promised. They promised to protect our rights, and they have not done that."[18]

My second analogue to the upstream temple god arises from the combination of the urgent danger that all human beings face and the circumstances of common fate this danger creates for us. Urgent danger is a fabulous catalyst for action. "Depend upon it, Sir, when a man knows he is to be hanged in a fortnight," wrote the great English essayist Samuel Johnson, "it concentrates his mind wonderfully."[19] But while the urgency of our situation may be starting to dawn on many people, the reality that we're in a situation of common fate—that the danger we face is an implacable equalizer—hasn't fully done so yet.

Many people still think they can escape. They ask: "Where can we go?"—as in, "Where can we go to be safe when things start falling apart?" Some people speculate about going "north," to flee the heat and migrating masses; others fantasize about going to Mars; zillionaires are buying estates with private airstrips in New Zealand, so they can scurry to that distant land when the going gets rough.

Yet the problems we're collectively creating for ourselves are ramifying through every component of our global socio-ecological system, so no one can escape or hide from their consequences, and no place on Earth will remain unscathed. (Very few will get to Mars in the foreseeable future.) Being at the end of the lifeboat farthest from the leaks doesn't mean winning the game; it just means having more time to observe the horrible processes of loss before one is engulfed in turn.

In humanity's past, the combination of urgent danger and common fate arose most clearly in situations of war. Societies mobilizing against an external enemy were often able to create a strong "we" identity in the process. Tolkien exploits our collective memory of such situations by imagining for his Fellowship of the Ring an external, personified enemy (Sauron) that threatens the group with annihilation. Today, alas, as Walt Kelly has pithily observed, the enemy is us, so no external enemy is readily available to catalyze the "we" identity we desperately need.

But perhaps an external *personified* enemy isn't necessary. Instead, we need to recognize our common human vulnerability and then clearly identify for ourselves superordinate goals—goals that can't be achieved unless we all cooperate—that address our vulnerability.

This seems to be exactly what the subaks did in Bali. They recognized their common vulnerability to shortages of irrigation water, and then they identified for themselves a clear superordinate goal to prevent those shortages: appeasing the upstream temple god. Within their worldview, they believed that goal could only be achieved by worshipping the god *together*. The result was a strong "we" identity—an instance of what I call "a common-fate community"—and, as vital side benefit, the remarkable cooperation that has allowed them to govern their water system.

## COMPLEMENTARY IDENTITIES

The notion of our common humanity—that all human beings share the same fundamental moral worth—originated in the Axial Age. It was then, in the millennium before the Common Era, says the philosopher Karl Jaspers, that the idea of "the human being as we still conceive of it today made its appearance."[20] Siep Stuurman, a Dutch intellectual historian, notes that the founders of the Axial Age usually held "strongly hierarchical views of social order" but at the same time "embraced inclusive visions of humanity as an overarching moral community." Today, the idea of our common humanity is codified in agreements like the 1948 Universal Declaration of Human Rights.

The related but distinct notion of our common human identity—akin to a species-wide "we"—also arose long ago. "I am a citizen of the world," famously declared Diogenes, the ancient Greek philosopher and Cynic in the fourth century BCE. While the idea of our common humanity is now seen as commonplace, the idea of a common human identity is far less widely accepted. Still, today it appears in such different worldviews as the Bahá'í faith and the South African humanist philosophy of Ubuntuism, which emphasizes the connectedness of all humanity (and by some accounts, all living beings). It's also institutionally embodied in organizations like Médecins Sans Frontières (Doctors without Borders) and the World Federalist Movement, which advocates for the creation of a global federal system of democratic states.

The two ideas together—that all human beings share a moral community and also a common identity—are now usually referred to as "cosmopolitanism." In recent years, critics of cosmopolitanism have been vicious. Some, like the British political philosopher John Gray, attack the very concept of "humanity" itself. "If you strip away religion and metaphysics and think of the human species in strictly naturalistic terms," he declares, "you will see that 'humanity' . . . is a figment of the imagination." "On the conflict-ridden earth," he continues, "human beings are raucously diverse and often savagely divided in their values."[21] Others disparage the notion of global citizenship. "If you believe you're a citizen of the world," former British prime minister Theresa May notoriously announced, "you're a citizen of nowhere."[22] For these critics, who are usually nationalistic and ideologically conservative, robust personal and group identities arise from people's intimate connections with their communities and nations, and with the histories, traditions, habits of mutual aid and civic obligation, and transcendent values that these communities and nations transmit. Cosmopolitanism, in their view, strips people out of this context of social solidarity and risks producing rootless, anomic, and ultimately violent individuals.

But these critics set up a straw man. There's no reason in principle why our respective community and national "we" identities can't complement, in our minds, our global "we" identity. There's no reason in principle why, depending on our needs and social contexts, we can't see ourselves as members of our respective linguistic, faith, and national communities, for instance, at the same time as we identify with humanity as a whole. Such pluralism of intersecting identities across scales from the local to the global is something to be cultivated and treasured; it can make for vibrant and resilient societies.[23]

Our "we" identities are, as Gray says, figments of our imagination, but I'd argue that's a good thing. It means we can reimagine and reconstruct them in ways that might change our behavior and allow us to flourish together on Earth. The state-space and binding questions help us see some ways this could be possible, particularly, as I mentioned earlier, the questions on Social Differentiation, Source of Personal Identity, and the Relationship between Humans and Nature.

**Social Differentiation**

First, if our positive alternative worldview is to include at its core the idea of an encompassing global "we," it cannot see the differences between groups of people around the world as large and essential—in other words, it cannot emphasize humanity's essential social differentiation. Instead, while we beneficially and accurately acknowledge that people have a plurality of identities, we need to emphasize the overriding fact of people's basic similarities.[24] If there's one thing that distinguishes highly conflict-prone worldviews—those looking for a fight—from worldviews that aren't, it's their respective positions on this question of essential social differentiation. It will be the most important distinction between the Mad Max worldviews, which will emerge from the disruptions this century, and the positive alternatives we must create to counter them.

**Source of Personal Identity**

But this belief in people's basic similarity can't remain just an intellectual proposition. We must each take it on board emotionally, which means we must each make personal membership in the global "we" part of our individual identity. When we ask ourselves "Who am I?" our answer should include, in part, the response "I am a member of the human community."

**Relationship between Humans and Nature**

And finally, it's important to acknowledge that there's no clear boundary between human beings and their surrounding natural world—that the natural world is intimately part of us. This means that the boundary of our identity—of our "we"—must expand to encompass nature too. "To regain our full humanity," writes the systems theorist Fritjof Capra, "we have to regain our experience of connectedness with the entire web of life."[25] Or as Daniel Wildcat, a professor at Haskell Indian Nations University in Kansas says: "Think of how our worldview changes if we shift from thinking that we live in a world full of resources to a world where we live among relatives."[26] The still widespread belief in human separateness from nature, and in human exceptionalism, made it easier for us to devastate Earth's biosphere. And now the rebound effects are becoming catastrophic.

THE ABOVE THREE IDENTITY COMMITMENTS will affect the scope of our commitments to justice. We see people (and other living things) who are part of our "we" group as within range of our responsibility and care. So, the broader our "we," the broader the reach of our group's principles of justice and the protection they afford.[27] In such a moral community, we assume that the actions, interests, and values of any given member are morally legitimate—that is, right and just—unless we learn that they've grossly violated the community's principles in some way.

And the relationship between our identity and justice commitments operates in the other direction, too: if we don't act to enhance justice, we're going to find it much harder to create a global "we" identity. Enormous and growing inequalities in income, wealth, status, and power in and between our societies—across gender, racial, ethnic, religious, and national divides, for instance—will stymie any and all efforts to create a common human identity, if they're left unaddressed. That's another reason why, when it comes to climate change, the principle of "common and differentiated responsibilities" is so important; by making the burden of costs and responsibilities for dealing with the problem fairer across nations, it encourages all humanity to pull together to solve the problem.

### OUR IMMORTALITY PROJECT

Powerful hope is hope that motivates us to work together as a species to solve our common problems, and to win, if necessary, any future struggle with Mad Max foes. We'll have powerful hope if we have a compelling vision of the future that's supported by appropriate worldviews. I've just sketched four principles—commitments to opportunity, safety, justice, and a shared global identity—that I believe must be at the core of this vision. They can provide the cognitive and normative foundations for strong collective action around the planet—through governments, transnational organizations, and global civil society—to solve our problems.

But we need something else, too. We need an "immortality project" for all humanity, one that gives people and groups around the world broad possibilities to imagine, tell, and weave together their own hero stories—and to live them together—as we move towards a shared vision of the future. We need a project that, by uniting us, eases our mounting fear and death anxiety. What kind of superordinate goal could bring us together in this way—could bring together into a common "we" the incredibly diverse peoples on Earth: street sellers in Lagos, ranchers in Queensland, rickshaw wallahs in Varanasi, office staff in Shanghai, tech entrepreneurs in San Francisco, and tree-nut harvesters in the Amazon? What goal, in other words, could work for humanity today the way the goal of appeasing the upstream temple god worked for the subaks in Bali?

Our collective, species-wide immortality project is simple to describe, but it's going to be staggeringly hard to execute: not only must we must stop our collective slide towards global calamity—we need to reverse it. Simply stopping the slide—just keeping things from getting worse—isn't enough. Reversing it means aggressively addressing our world's agonizing social and economic injustices; and most importantly, it means *rebuilding nature*. We must commit to fixing the hideous environmental mess we've made—with all the science, technology, economic investment, and intellectual, artistic, and emotional creativity we can muster. Crucially, this doesn't mean returning nature to its state prior to human domination of Earth—a sort of *status quo ante*. It means, instead, accepting that nearly all the species that humans have extinguished and ecosystems we've wrecked are gone forever, but also recognizing simultaneously that with our intelligence and vital imagination we can create the conditions for nature's complexity to bloom in new ways across the planet's surface—not just in remaining unpopulated zones, but in our densest cities too. We can help life on Earth fully express once more its profound creativity, explore anew its vast adjacent possible, and recreate its magical web of complexity around us—perhaps by accelerating biological processes of evolution.

Only a truly audacious commitment of this kind could generate within us and our societies the moral passion and emotional excitement—the combination of joy, awe, and compassion—that appeals to the exuberant, prudent, and empathetic temperaments I've identified and provides us the powerful motivation to persevere through the difficult times ahead.

The motivation will derive partly from an explicitly spiritual connection with the sacredness of life and nature. "On this planet where astrobiologists detect no other life in the galaxy," writes the American ecologist Carl Safina, "the rarity and perhaps even uniqueness of life in the universe makes Earth a sacred place. All known meaning in the universe is generated here, because this is the only living planet."[28] As Stuart Kauffman, the complexity scientist, has said: "How much magic do you need? This is 'God' enough for me."[29]

I bring all these elements together in the answers to the state-space questions on page 306, in what I call the "Renew the Future" worldview. Each of the four core principles we've considered—the commitments to opportunity, safety, justice, and to an identity that encompasses not only all peoples on the planet, but all life too—is reflected in this worldview's answers to the questions.[30]

The Renew the Future perspective also integrates reason and feeling as equally vital sources of understanding that can help us address our problems, while creating space for a moral and spiritual appreciation of our larger purpose amidst the complexity of life, nature, and the cosmos.[31] And it's oriented with hope towards the future and the considerable possibility of positive change.[32]

I've put the Mad Max worldview's answers to the same questions on the adjacent page, so you can compare the two perspectives easily.

**"RENEW THE FUTURE" STATE-SPACE ANSWERS**

| | ISSUE | QUESTION | BELIEF STRENGTH | | | | | | |
|---|---|---|---|---|---|---|---|---|---|
| IS | *Threat* | Is the world a safe or dangerous place? | **SAFE** | S | M | Ambivalent, no position | M | S | **DANGEROUS** |
| | *Source of Understanding* | Is the world best understood by using reason and feeling together or separately? | **TOGETHER** | S | M | Ambivalent, no position | M | S | **SEPARATELY** |
| | *Spirituality* | Is the world infused with a spirit? | **MATERIAL** | S | M | Ambivalent, no position | M | S | **SPIRITUAL** |
| | *Moral Principles* | Are moral principles subjective or objective? (Relativism vs. absolutism) | **SUBJECTIVE, CONTEXTUAL** | S | M | Ambivalent, no position | M | S | **OBJECTIVE, UNIVERSAL** |
| | *Agency* | Is a person's fate a result of circumstances or choice? (Determinism vs. free will) | **CIRCUM-STANCES** | S | M | Ambivalent, no position | M | S | **CHOICE** |
| | *Human Nature* | Are people basically generous or selfish? | **GENEROUS** | S | M | Ambivalent, no position | M | S | **SELFISH** |
| | *Relationship between Humans and Nature* | Are human beings as one with nature or distinct and exceptional? | **AS ONE WITH NATURE** | S | M | Ambivalent, no position | M | S | **DISTINCT AND EXCEPTIONAL** |
| | *Social Differentiation* | Are the differences between groups small and unimportant or large and essential? | **SMALL AND UNIMPORTANT** | S | M | Ambivalent, no position | M | S | **LARGE AND ESSENTIAL** |
| | *Source of Personal Identity* | Does one's identity derive mainly from oneself or from one's group? | **FROM ONESELF** | S | M | Ambivalent, no position | M | S | **FROM ONE'S GROUP** |
| OUGHT | *Time* | For inspiration, should one look to the future or to the past? | **TO THE FUTURE** | S | M | Ambivalent, no position | M | S | **TO THE PAST** |
| | *Change* | Should change be encouraged or resisted? | **ENCOURAGED** | S | M | Ambivalent, no position | M | S | **RESISTED** |
| | *Care for Others* | How much should one help others? | **A LOT** | S | M | Ambivalent, no position | M | S | **NOT MUCH** |
| | *Authority* | All things being equal, should one resist authority or defer to it? | **RESIST** | S | M | Ambivalent, no position | M | S | **DEFER** |
| | *Power* | Is the use of power over others usually wrong or often right? | **USUALLY WRONG** | S | M | Ambivalent, no position | M | S | **OFTEN RIGHT** |
| | *Wealth* | Are large differences in wealth immoral or moral? | **IMMORAL** | S | M | Ambivalent, no position | M | S | **MORAL** |

**"MAD MAX" STATE-SPACE ANSWERS**

| | ISSUE | QUESTION | BELIEF STRENGTH | | | | | | |
|---|---|---|---|---|---|---|---|---|---|
| IS | *Threat* | Is the world a safe or dangerous place? | **SAFE** | S | M | Ambivalent, no position | M | S | **DANGEROUS** |
| IS | *Source of Understanding* | Is the world best understood through reason or feeling (emotion/intuition)? | **REASON** | S | M | Ambivalent, no position | M | S | **FEELING** |
| IS | *Spirituality* | Is the world infused with a spirit? | **MATERIAL** | S | M | Ambivalent, no position | M | S | **SPIRITUAL** |
| IS | *Moral Principles* | Are moral principles subjective or objective? (Relativism vs. absolutism) | **SUBJECTIVE, CONTEXTUAL** | S | M | Ambivalent, no position | M | S | **OBJECTIVE, UNIVERSAL** |
| IS | *Agency* | Is a person's fate a result of circumstances or choice? (Determinism vs. free will) | **CIRCUM-STANCES** | S | M | Ambivalent, no position | M | S | **CHOICE** |
| IS | *Human Nature* | Are people basically generous or selfish? | **GENEROUS** | S | M | Ambivalent, no position | M | S | **SELFISH** |
| IS | *Relationship between Humans and Nature* | Are human beings as one with nature or distinct and exceptional? | **AS ONE WITH NATURE** | S | M | Ambivalent, no position | M | S | **DISTINCT AND EXCEPTIONAL** |
| IS | *Social Differentiation* | Are the differences between groups small and unimportant or large and essential? | **SMALL AND UNIMPORTANT** | S | M | Ambivalent, no position | M | S | **LARGE AND ESSENTIAL** |
| IS | *Source of Personal Identity* | Does one's identity derive mainly from oneself or from one's group? | **FROM ONESELF** | S | M | Ambivalent, no position | M | S | **FROM ONE'S GROUP** |
| OUGHT | *Time* | For inspiration, should one look to the future or to the past? | **TO THE FUTURE** | S | M | Ambivalent, no position | M | S | **TO THE PAST** |
| OUGHT | *Change* | Should change be encouraged or resisted? | **ENCOURAGED** | S | M | Ambivalent, no position | M | S | **RESISTED** |
| OUGHT | *Care for Others* | How much should one help others? | **A LOT** | S | M | Ambivalent, no position | M | S | **NOT MUCH** |
| OUGHT | *Authority* | All things being equal, should one resist authority or defer to it? | **RESIST** | S | M | Ambivalent, no position | M | S | **DEFER** |
| OUGHT | *Power* | Is the use of power over others usually wrong or often right? | **USUALLY WRONG** | S | M | Ambivalent, no position | M | S | **OFTEN RIGHT** |
| OUGHT | *Wealth* | Are large differences in wealth immoral or moral? | **IMMORAL** | S | M | Ambivalent, no position | M | S | **MORAL** |

They're radically different. Although both regard the world as a dangerous place, see one's identity as derived significantly from one's group, and regard the use of power over others as (at least sometimes) morally right, their answers are almost diametrically opposed on ten of the remaining twelve dimensions. These radically different perspectives will produce radically different social behaviors in any groups that adopt them; most obviously, the Renew the Future worldview will be far more conducive to cooperation, and collaborative problem-solving, across large groups of diverse people. And because these worldviews support starkly distinct visions of the future, they'll also have radically different implications for our hope.

To show how, I've used the list of binding questions (chapter 18) to draw cognitive-affective maps of the way people holding these respective worldviews would likely feel about the future and about their own "we" group's relationship to that future. A person of the Mad Max persuasion would see the future negatively, because it would be associated with a host of emotionally negative ideas like decline, scarcity, loss, and danger, and with threats from enemies and dangerous outsiders. In turn, the person would see their "we" group as needing protection from this fearful future, with violence and authority as ways to defend borders, secure territory, and ultimately create safety for the "we" group from external threats.

In contrast, a person holding the Renew the Future worldview would see the future positively, because "the future" would link to many other emotionally positive concepts—including the principles of opportunity, safety, justice, and the global "we" identity that I've just outlined. The superordinate goal of rebuilding nature supports both the development of the "we" identity and the idea of a positive future. The positive future then justifies hope, which, in combination with danger, motivates agency.

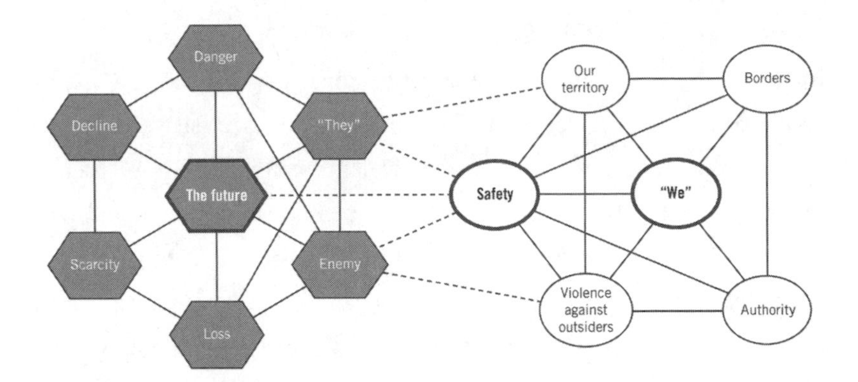


*A cognitive-affective map of the "Mad Max" vision of the future*

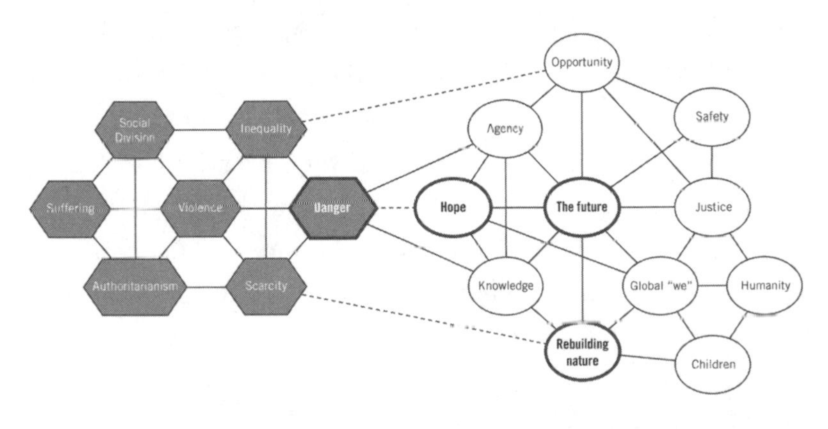


*A cognitive-affective map of the "Renew the Future" vision of the future*

The hope in the diagram above is the commanding hope I've argued for in this book—hope that's honest, astute, and powerful. It's honest, because it's informed by knowledge, and that knowledge in turn incorporates a true understanding of danger; in other words, it honestly acknowledges the reality of the unparalleled challenges we face. It's astute, because the knowledge that informs our hope also includes a deep understanding of others' worldviews, which boosts the strategic smarts needed for strong, just agency.

Most importantly, this hope interposes itself between the idea of danger and the idea of the future by helping to counteract danger's negative emotions. The dotted line between "Danger" and "Hope" indicates the two concepts are emotionally antithetical: more negative emotion associated with danger means less positive emotion from hope; while more positive emotion associated with hope means less negative emotion from danger.[33] Commanding hope recognizes the possibility of the negative outcomes on the diagram's left, but these outcomes aren't seen as inevitable. Without succumbing to delusion and dishonesty, this hope keeps those negative emotions from penetrating into and infecting the vision of the future on the diagram's right. In this way, that vision can continue to sustain the powerful agency that can help make the vision real.

Is this a realistic starting point for the vision of the future we so urgently need? Can enough people around the world feasibly cross the mindscape to reach a widening worldview basin of this kind, filling that basin with a strong cosmopolitan concept of "we" identity, commitments to opportunity, safety, and justice, and the superordinate goal of rebuilding nature? Can it be the basis for a collective, commanding hope?

The mindscape's geography today looks forbidding. Today's worldview basins seem enormous, deepened by people's fears of meaninglessness, by dense, interdependent networks of locked-in institutions and technologies, and by countless powerful vested interests. Yet these basins won't stay this way. Events are going to get very hot. Emotions will surge in waves around the planet; social moods will flip one way and then flip another, and then flip again. Today's politics, alliances, routines, and patterns of thinking will become radically outdated and out of place. Current structures of power will crack and splinter. Under these conditions, could we leap across the mindscape? Could a large part of humanity jump to a broad new basin that harbors positive worldviews of the character I've described?

In a highly connected and white-hot world, I'm convinced such a jump is *feasible*. Hundreds of millions and maybe even billions of people could quickly converge around principles like those I've just discussed.

But could such a jump, assuming it happens, be *enough* to genuinely reduce the danger humanity faces? That's not so clear. Whether it's enough will depend on whether the worldview change catalyzes the necessary transformation of our institutions and technologies, particularly those that underpin our current carbon-based energy system and our calamitous model of economic growth. It will depend, too, on whether enormous, irreversible, and harmful shifts in climate, food systems, economies and the like have already taken place—whether, in other words, the worldview jump happens just soon enough to avert those irreversible shifts or too late to do so.

And, finally, whether the jump is ultimately *enough* will depend greatly on the outcome of the emerging battle with other, hostile worldviews. But better to wage this fight together (a heroic fight, if ever there were one) for humanity's ideational and moral future and then lose it, than to concede the mindscape to such a vile end without a fight at all.

# Epilogue

# The Battle for Tomorrow

*All of us . . . are preparing a renaissance beyond the limits of nihilism. But few of us know it.* Albert Camus

IN 2019, THE BATTLE BEGAN in earnest. Activists around the world—particularly young people galvanized by the example of their own peers, like Greta Thunberg, the Swedish sixteen-year-old who started the School Strike movement—took to the streets by the millions in the majority of the world's countries, demanding real action on climate change. The youths leading the School Strike, Extinction Rebellion, and (in the United States) Sunrise movements rightly defined the issue as one of survival, not just for themselves later this century, but for a large portion of life on the planet. And with their moral clarity, mass engagement, and unalloyed bravery, they articulated a vision of hope for the future unlike any offered in response to this crisis to date.

As I write these words in mid-2020, the COVID-19 pandemic has abruptly pushed all other issues, including the climate crisis, into the background in the mainstream media. But the battle for tomorrow still continues along countless hidden fronts, and it will return to the foreground once the pandemic's immediate economic disruptions begin to recede and impediments to protest, like social distancing, are removed. That's because the climate crisis and the social harms and injustices it causes aren't going away. Their dreadful implications for everyone's future will only become relentlessly more obvious—and terrifying.

The pandemic's vast human toll in pain, grief, anxiety, and loss is merely a glimpse of what our species can expect if we continue headlong down our current path. It has also given us a vivid example of a global nonlinearity, a cascading sequence of changes in which multiple social systems flip simultaneously to distinctly new states; and it's a signal that humanity's planet-spanning economic, social and technological networks are critically vulnerable and unstable. If, in the pandemic's wake, humanity returns to business as usual, we can expect ever more frequent global nonlinearities—social earthquakes—of ever higher destructive force.

But cascading changes in our global social systems don't have to be so pernicious. By reminding us of our common fate on this tiny planet, the pandemic could catalyze an urgently needed shift in humanity's collective moral values, priorities, and sense of self and community, along the lines of the Renew the Future worldview I described in the last chapter. COVID-19 is a collective problem that requires global collective action—just like climate change—and we won't address either challenge effectively if we retreat into our narrow identities and wall ourselves off from each other.

We can harbor honest hope for such a positive normative shift, but astute hope requires that those of us, young and old, battling for a better future keep clearly in mind what we're up against. We're confronting some of the most formidable vested interests on Earth—coal, oil, and natural gas companies; the associated pipeline, refining, and tanker industries; and the banks, investors, and stockholders that finance them, including, through their pension and mutual funds, countless people not directly associated with the fossil-fuel sector. And then there's a vast penumbra of industries, sectors, and groups tightly interwoven with the fossil-fuel energy system—including modern industrial agriculture; most electricity utilities, especially in poorer countries; and much of the transport sector, particularly the rail, trucking, and airline industries.

Those who see the climate struggle ahead for what it is are rightly calling for the replacement of the encompassing, entrenched, powerfully interlinked set of worldviews, institutions, and technologies in which we're all currently embedded. Their opponents are often responding with implacable ideological hostility and scurrilous attacks on activist leaders. Around the world they're also manipulating or capturing state power in a rear-guard effort to slow change to the dominant energy system, attacking climate science and scientists, and launching the kind of McCarthyesque inquisitions against environmentalists that anti-nuclear campaigners experienced in Stephanie May's day. And all of this is happening even before things get really bad—before the demographic, climate, economic, and other stresses that are building under the surface of our global systems start causing pulses of devastating social earthquakes.

The closest historical analogue to our current situation—in terms of the nature and scope of the change humanity needs to make and the political struggle required—is probably the worldwide fight against slavery from the eighteenth to the twentieth centuries, as has been noted by the historian Jean-François Mouhot and author and activist Naomi Klein, among others.[1] The analogy is far from perfect, and it's hugely contentious. Slavery was a unique moral abomination. Among other features, it hinged on the institutional dehumanization of the enslaved, and the harm it caused people was direct and intentional. Fossil-fuel production and use don't depend on *institutionalized* dehumanization, nor are the harms these practices cause to people (through climate change) direct or intentional.

Still, our fossil-fuel regime is fundamentally an energy system that powers the world economy, like slavery was in its day, and this regime has a similar multiplicity of entanglements with every facet of the societies in which it thrives. Also, climate change is inescapably a moral challenge, like slavery was, particularly as the harm falls most severely on marginalized human beings. And then there's the disturbing truth noted by several scholars, such as the American environmental thinker David Orr, that the arguments used to defend the fossil-fuel regime today bear striking resemblance to those defenders of slavery once used. For example, defenders of slavery said further cultural and economic advancement depended on the slave system, while defenders of the fossil-fuel regime today say economic growth and human well-being around the planet depend on continued, and even increasing, consumption of carbon-based energy.

The struggle to end slavery hints at what's required now and what's in store in the coming years if humanity is to find a path through its manifold crises to a halfway decent future. It suggests that activists trying to produce change must be aware, first and foremost, of their relative weakness—in terms of conventional political and economic power—compared to their opponents. This power imbalance is made far worse by the breakdown of democratic processes across much of the world, accompanied by the rise of populist authoritarianism, and, more fundamentally, by the breakdown in many democracies of the shared understanding of "facts" and commitment to the commonweal that are essential for effective democratic deliberation.

All this means that we must look for ways to use the dominant system of worldviews, institutions, and technologies against itself, by exploiting—a bit like a jujitsu fighter—its energies, weaknesses, and leverage points, so as to produce dramatic nonlinear, tipping-point change—or what I've called in this book virtuous cascades of change. We can begin by multiplying today's vital youth activism a thousand-fold around the world (not inconceivable, given the mounting day-to-day evidence of the catastrophic effects of climate change), mobilize this activism into coherent political movements that genuinely challenge dominant power systems (a process that's likely to provoke a vicious and quite possibly violent reaction), and then let the propulsive mechanisms of the global financial system do the rest—as institutional and individual investors everywhere abruptly recognize that the days of the fossil-fuel economy are finally coming to an end, and it's time to flee the sinking ship.

Despite the daunting odds, there's ample reason for hope as we've conceived it here: a commanding hope that's honest about the dangers ahead and astute about the strategies we should use to face those dangers, and one that powerfully motivates us, as agents, to push through adversity and work together to create the future we want. It will take immense effort starting immediately, of course—including effort to develop, experiment with, and use such tools as our intelligent minds can create. But what could be more worthwhile?

Let's not aim for what's merely feasible and falsely hope it will be enough. Instead, with commanding hope, let's aim for what we'd *all* consider enough—a future in which our children and life on this planet can flourish—and then strive to make that future our reality.

# Acknowledgments

I COULDN'T HAVE WRITTEN ANYTHING resembling this book without the inspiration and emotional support that Ben and Kate, my children, have provided so abundantly over the last eight years. They've been with me either in spirit or in person every step of the way. Nor could I have written it without my wife and partner, Sarah, who has always been by my side with love, comfort, and guidance, encouraging me to press forward, most crucially after I finally decided (in the face of much advice) that this book had to be about hope.

I was away from my family at solo writing retreats for a total of nearly seven months. I missed them all terribly on these occasions; and every time I came home, I found that the kids had changed. I'll never get that time with my family back, and I'll be forever indebted to Sarah, Ben, and Kate for giving me those precious opportunities to focus exclusively on writing.

I'm grateful to my literary agent, Bruce Westwood, for gently urging me, in 2011, to write one more book. Over the twenty-five years we've known each other, I've been blessed to become friends with Bruce, to be touched by his warmth and charm, and to benefit from his enormous professional acumen. Louise Dennys, my editor in Canada, and Jonathan Cobb, my editor in the United States, helped turn a heap of somewhat inchoate ideas and arguments into a disciplined and (I trust) compelling book. Jonathan's wonderful insights and unflagging patience were critical at moments when I felt lost; sometimes he seemed to have a clearer sense for the essence of my ideas than I did myself. Louise's brilliant and painstaking attention to every aspect of the book through multiple revisions has crucially strengthened its argument and tightened and enlivened its story. To you both, thank you from the bottom of my heart.

Geoffrey and Elizabeth May were extraordinarily generous in introducing me to their mother's story and helping me understand its details and context. Stephanie's memoirs and scrapbooks are a goldmine of information about her activism and perspective on the world and about how life and politics looked and felt in the United States in the late 1950s. Geoffrey took me through Stephanie's old house in Cape Breton, Nova Scotia, regaling me room by room with family tales; and Elizabeth and I discussed the role of hope in today's environmental activism. Both graciously answered my many questions and read and commented on all passages about their mother.

At the University of Waterloo, my dean, Jean Andrey, has enthusiastically supported this book project for many years, even though it falls far outside the ambit of normal scholarship. My students and colleagues at the university and beyond have helped me develop and strengthen the scientific foundations of my argument, particularly the parts concerning complexity science, psychology, political science, and processes of institutional and technological change. Paul Thagard, Tobias Schröder, Manjana Milkoreit, and Jinelle Piereder pioneered our work on cognitive-affective maps. I've also learned much from Paul on the role of emotion in cognition, from Manjana on the role of imagination in envisioning new futures, from Mike Lawrence on dehumanization and on the tension between global diversity and uniformity, and from Jonathan Leader Maynard (at the University of Oxford) on the connections between political ideology and violence. Mike and Norman Kearney explored with me the implications of the WIT framework. Steven Mock advanced my thinking about symbols of nationalism and group identity, Steve Quilley about the ontological and moral commitments of modern conservatism, Lee Smolin about the nature of time, and Stephen Purdey about the importance of new, emotionally grounded, global narratives for humanity. Tahnee Prior explained the Finnish concept of *sisu*. Yonatan Strauch deepened my grasp of the complexity of energy transitions and Clay Dasilva of our societies' psychological and social commitments to economic growth.

Scott Janzwood gave careful attention to my claims about uncertainty and wrote excellent research reports for me on normative cascades, the neurobiology of hope, humanity's mood shift since 2000, and the parallels between the fights against slavery and climate change.

I owe special debts to my dear friends Frances Westley, for her groundbreaking research on social innovation, and Fred Bird, for his advice on the Axial Age, Karl Jaspers, Reinhold Niebuhr, Hannah Arendt, Émile Durkheim, and the history of hope (the subject of his own book project). Matto Mildenberger (University of California, Santa Barbara) straightened out my ideas on Pascal's Wager and on the politics of climate change in advanced economies. Sheldon Solomon (Skidmore College), a world authority on Ernest Becker and Terror Management Theory, reviewed and commented on the passages dealing with those topics. Jordan Mansell (University of Quebec at Montreal) completed initial surveys and tests of the state-space model of political ideology.

Anthony Barnosky and Elizabeth Hadly (Stanford University) reviewed the passage on their co-authored article about planetary state shifts, which Kate found on Sarah's desk; John Bongaarts (Population Council) checked my assertions regarding African demographic momentum; Gordon Laxer (formerly of the Parkland

Institute) contributed his thoughts on the links between perceptions of justice and identity; Stuart Kauffman (Santa Fe Institute) opened my mind to the possibilities of the sacred in living nature's complexity; and James Risbey (Australia's Commonwealth Scientific and Industrial Research Organization) reviewed the graph in chapter 2 on warming since the beginning of the Holocene epoch. The concept of "virtuous cascades" originated with Trevor Hancock (University of Victoria). Jane Long (California Council on Science & Technology) shared with me her ideas about our responsibilities to sustain others' hope, and Jim Balsillie helped me understand the pernicious role of social media giants in eroding social solidarity.

Many thanks, too, to Joan Hewer, for her tireless assistance with research and locating articles and books over the years and for handling all illustration permissions; Jacob Buurma for his beautiful illustrations; David Porreca (University of Waterloo) and Rabun Taylor (University of Austin, Texas) for guidance on the book's Latin dedication; Nick Garrison, Jane Willms, and John Pearce for discerning comments on the book's opening (and also to Jane for her company and insights during our travels so many years ago); Werner Kurz and Sarah Beukema for general solace during my writing retreats and for reviewing my calculations on the volume of carbon dioxide that needs to be removed from Earth's atmosphere to keep warming to 1.5 degrees Celsius; and Matt Hoffmann (University of Toronto) for his heartfelt comments on social scientists' fear of climate change.

This book has benefited immeasurably from the keen editorial eye of Associate Editor Rick Meier (Knopf Canada), who saw the book through to production; from the meticulous care of my empathetic copyeditor, Linda Pruessen, and proofreader, Tilman Lewis; and from the patience and skill of Senior Managing Editor Deirdre Molina, who brought the book to completion. I've also been enormously fortunate, once again, to have had financial support from the Winslow Foundation to defray research, illustration, and editing costs.

And, finally, a special tip of the hat to the blue monkeys from the land of Twang, a tortoise named Bill, and their cats—all creations of the British comedian Eddie Izzard, who has brought so much laughter to our family and who somehow manages to make the world's craziness a little more tolerable.

# A Note on the Book's Dedication

"*Nobis non desistendum est*" is Latin for "We must not give up." The verb *desistere* means "to dissociate oneself," "to give up, leave," or "to stop, cease, desist (from)." The form here is gerundive (future passive participle, acting as a verbal adjective), which in Latin denotes something that should or must be done. Cato's famous declaration "*Carthago delenda est*" had the same formulation. The *non* is the negatory, and *nobis* expresses in the dative plural the agent "we" of the gerundive's action. I use "we," because the dedication is addressed directly to my children, and indirectly to all of us in the future, as we stand together to face its trials.

# Endnotes

## PROLOGUE: OUR OWN STORY

1. Anthony D. Barnosky, Elizabeth A. Hadly, Jordi Bascompte, Eric L. Berlow, James H. Brown, Mikael Fortelius, Wayne M. Gertz, John Harte, Alan Hastings, Pablo A. Marquet, Neo D. Martinez, Arne Mooers, Peter Roopnarine, Geerat Vermeij, John W. Williams, Rosemary Gillespie, Justin Kitzes, Charles Marshall, Nicholas Matzke, David P. Mindell, Eloy Revilla, and Adam B. Smith, "Approaching a State Shift in Earth's Biosphere," *Nature* 486 (June 2012): 52–58, doi: 10.1038/nature11018.
2. Thomas Homer-Dixon, *The Ingenuity Gap: How Can We Solve the Problems of the Future* (New York: Knopf: 2000); and Homer-Dixon, *The Upside of Down: Catastrophe, Creativity, and the Renewal of Civilization* (Toronto: Knopf Canada, 2006).
3. On how our stories influence our destiny, see David C. Korten, *Change the Story, Change the Future: A Living Economy for a Living Earth* (Oakland, CA: Berrett-Koehler Publishers, 2015); and Martin Puchner, *The Written World: The Power of Stories to Shape People, History, Civilization* (New York: Random House, 2017). Yuval Noah Harari offers a more skeptical view of the value of stories in his deeply insightful book *21 Lessons for the 21st Century* (New York: Spiegel & Grau, 2018), especially in chapter 20, "Meaning: Life Is Not a Story," 275–315.
4. Jonathan Lear, *Radical Hope: Ethics in the Face of Cultural Devastation* (Cambridge, MA: Harvard University Press, 2006); Joanna Macy and Chris Johnstone, *Active Hope: How to Face the Mess We're in*

*without Going Crazy* (Novato, CA: New World, 2012); and Kate Davies, *Intrinsic Hope: Living Courageously in Troubled Times* (Gabriola Island, BC: New Society, 2018).

## CHAPTER 1: SIGNALS

1. Hannah Arendt, *The Origins of Totalitarianism* (Cleveland: Meridian, 1958 [1951]), vii.
2. Jack Schubert and Ralph Lapp, "Radiation Dangers: What Is Known about the Hazards of Bomb Tests? What Risks Is It 'Reasonable' to Run?", *New Republic* (May 20, 1957), 10.
3. United Nations Scientific Committee on the Effects of Atomic Radiation (UNSCEAR), *Sources and Effects of Ionizing Radiation, 2000 Report, Volume I: Report to the General Assembly, with Scientific Annexes—Sources* (New York: United Nations, 2000), 5, doi: https://doi.org/10.18356/a2ed0f47-en.
4. Elizabeth May, "Chapter 1: Democracy 101," *How to Save the World in Your Spare Time* (Toronto: Key Porter, 2007), 14.
5. Three years earlier, in 1959, twelve countries, including the nuclear powers at the time, had signed the Antarctica Treaty. It banned military activities and established freedom of scientific research on the continent.
6. The smoke came again, worse than ever, in the summer of 2018. Of course, fire is a natural part of ecological change in forests, and enormous fires producing prodigious quantities of smoke have happened often in recorded history in North America—most notably the great Chinchaga River fire in the Peace River district of British Columbia and Alberta in 1950. The smoke drifted southeastward across southern Ontario and on to New York, Washington, DC, and as far as Florida; it was so thick that motorists in Toronto had to turn on their lights in midday. But today's wildfires are on average far more frequent, intense, and extensive. In living memory, until the 2015, 2017, and 2018 events, smoke from fires on the mainland was never dense or extensive enough to cover Vancouver Island.
7. Vivian Wang, Donald G. McNeil Jr., Farnaz Fassihi and Steven Lee Myer, "Surge in Cases Raises Concern of a Pandemic: Virus Spreads in Iran and South Korea," *New York Times*, February 22, 2020.

8. Steven Pinker, *The Better Angels of Our Nature: Why Violence Has Declined* (New York: Penguin, 2012).
9. I've borrowed this phrase from John L. Casti and his provocative book *Mood Matters: From Rising Skirt Length to the Collapse of World Powers* (New York: Copernicus, 2010).
10. In a speech to the 2019 Conservative Political Action Conference, US president Donald Trump said: "Every day, we're restoring common sense and the timeless values that unite us all. We believe in the Constitution and the rule of law. We believe in the First Amendment right. And we believe in religious liberty. And we believe strongly in the Second Amendment and the right to keep and bear arms, which is under siege, folks. They have a lot of plans. It's under siege. Be careful. But I'll protect you, I promise you that. I'll protect you." Donald J. Trump, "Remarks by President Trump at the 2019 Conservative Political Action Conference," March 2, 2019, https://www.whitehouse.gov/briefings-statements/remarks-president-trump-2019-conservative-political-action-conference/.

## CHAPTER 2: HOW ABOUT . . .

1. Cognitive scientists Alison Gopnik and her colleagues have found that learners become less flexible as they transition from childhood to adulthood: "they are less likely to adopt an initially unfamiliar hypothesis that is consistent with new evidence. Instead, learners prefer a familiar hypothesis that is less consistent with the evidence." See Alison Gopnik, Shaun O'Grady, Christopher G. Lucas, Thomas L. Griffiths, Adrienne Wente, Sophie Bridgers, Rosie Aboody, Hoki Fung, and Ronald E. Dahl, "Changes in Cognitive Flexibility and Hypothesis Search across Human Life History from Childhood to Adolescence to Adulthood," *Proceedings of the National Academy of Sciences* 114, no. 30 (July 25, 2017): 7892–99, doi: 10.1073/pnas.1700811114.
2. Energy systems analyst Zeke Hausfather, in an assessment of past warming predictions for the climate-science website Carbon Brief, writes: "Climate models published since 1973 have generally been quite skillful in projecting future warming. While some were too low and some too high, they all show outcomes reasonably close to

what has actually occurred, especially when discrepancies between predicted and actual $CO_2$ concentrations and other climate forcings are taken into account." In 1981, for instance, James Hansen and his colleagues published "an estimate of early-2000s warming close to observed values." See Zeke Hausfather, "Analysis: How Well Have Climate Models Projected Global Warming?" Carbon Brief, October 5, 2017, https://www.carbonbrief.org/analysis-how-well-have-climate-models-projected-global-warming. While climate scientists have been largely right about general warming trends, they still can't precisely predict how warming will affect weather patterns at the regional and local level. Some now argue that a decline in Arctic sea ice has weakened the polar vortex winds and changed the behavior of the northern jet stream, in turn occasionally letting large quantities of extremely cold air escape from the Arctic basin into temperate latitudes, with warm air simultaneously flooding into the Arctic from the south. This mechanism may explain the alternating episodes of record warmth and cold during recent North American winters. See M. Kretschmer, D. Coumou, E. Tziperman, and J. Cohen, "More-Persistent Weak Stratospheric Polar Vortex Linked to Cold Extremes," *Bulletin of the American Meteorological Society* 99 (2018): 49–60; and D. Coumou, G. Di Capua, S. Vavrus, L. Wang, and S. Wang, "The Influence of Arctic Amplification on Mid-Latitude Summer Circulation," *Nature Communication* 9, no. 2959 (2018), doi: 10.1038/s41467-018-05256-8.

3. The extra energy trapped in Earth's atmosphere, at 0.75 watt per square meter over Earth's surface area of roughly five hundred trillion square meters, amounts to about thirty-two million terajoules (trillion joules) each day. The Hiroshima bomb released about sixty-three terajoules of energy, or about one five-hundred-thousandth as much.
4. Stefan Rahmstorf, "Paleoclimate: The End of the Holocene," RealClimate: Climate Science from Climate Scientists, September 16, 2013, http://www.realclimate.org/index.php/archives/2013/09/paleoclimate-the-end-of-the-holocene/.
5. Homer-Dixon, *The Upside of Down*, 278.

## CHAPTER 3: FIGHTING A SCARCITY OF HOPE

1. Donella H. Meadows, Dennis L. Meadows, Jørgen Randers, William W. Behrens III, *The Limits to Growth: A Report for the Club of Rome's Project on the Predicament of Mankind* (New York: Universe Books, 1972).
2. The American science writer Brian Hayes offers a thoughtful critique in "Computation and Human Predicament: The *Limits to Growth* and the Limits to Computer Modeling," *American Scientist* (May–June 2012): 186, doi: 10.1511/2012.96.186.
3. Graham Turner, *Is Global Collapse Imminent: An Updated Comparison of The Limits to Growth with Historical Data*, Research Paper no. 4 (Melbourne: Melbourne Sustainable Society Institute: August 2014).
4. For a technical explanation of why scarcity arising from the declining *quality* of mineral resources is not a binding economic constraint, see Magnus Ericsson, Johannes Drielsma, David Humphreys, Per Storm, and Pär Weihed, "Why Current Assessments of 'Future Efforts' Are No Basis for Establishing Policies on Material Use—a Response to Research on Ore Grades," *Mineral Economics* 32 (2019): 111–21.
5. Stanford University resource economists Marshall Burke and Vincent Tanutama estimate that "since 2000, warming has already cost both the US and the EU at least $4 trillion in lost output, and tropical countries are greater than 5% poorer than they would have been without this warming." They also note that "additional warming will exacerbate inequality, particularly across countries, and that economic development alone will be unlikely to reduce damages, as commonly hypothesized." Marshall Burke and Vincent Tanutama, "Climatic Constraints on Aggregate Economic Output," Working Paper no. 25779 (Cambridge, MA: National Bureau of Economic Research, 2019), http://www.nber.org/papers/w25779.
6. T. Schuur, "Permafrost and the Global Carbon Cycle," *Arctic Report Card: Update for 2019* (Silver Spring, MD: National Oceanic and Atmospheric Administration, 2019), https://arctic.noaa.gov/Report-Card/Report-Card-2019/ArtMID/7916/ArticleID/844/Permafrost-and-the-Global-Carbon-Cycle.

7. Franziska Gaupp, Jim Hall, Stefan Hochrainer-Stigler, and Simon Dadson, "Changing Risks of Simultaneous Global Breadbasket Failure," *Nature Climate Change* 10 (2020): 54–57, doi: 10.1038/s41558-019-0600-z.
8. The possibility of a "cascade of tipping points" in the global climate system and biosphere is highlighted in Timothy Lenton, Johan Rockström, Owen Gaffney, Stefan Rahmstorf, Katherine Richardson, Will Steffen, and Hans Joachim Schellnhuber, "Climate Tipping Points—Too Risky to Bet Against," *Nature* 575 (November 28, 2019): 592–95.
9. To some extent, this invention is already well underway, as we can see in the Transition Town movement for local economic autonomy, the rapid evolution of "green" political ideology in some Western societies, and various speculative, often science-fictional elaborations of alternative social orders, such as Ernest Callenbach's 1975 book *Ecotopia*.

## CHAPTER 4: SO OUR SOULS CAN BREATHE

1. Arendt, *Origins*, vii.
2. Anthony Leiserowitz, Edward Maibach, Seth Rosenthal, John Kotcher, Matthew Ballew, Matthew Goldberg, and Abel Gustafson, "2. Emotional Responses to Global Warming," in *Climate Change in the American Mind: December 2018*, Yale Program on Climate Change Communication, January 22, 2019, https://climatecommunication.yale.edu/publications/climate-change-in-the-american-mind-december-2018/4/.
3. T.S. Eliot, "Burnt Norton," *Four Quartets* (Boston: Houghton Mifflin Harcourt, 1943).
4. Charles R. Snyder, "Hope Theory: Rainbows in the Mind," *Psychological Inquiry* 13, no. 4 (2002): 258.
5. Gabriel Marcel, *Homo Viator: Introduction to a Metaphysic of Hope* (New York: Harper & Row, 1962), 10–11.

## CHAPTER 5: THE WAY HOPE WORKS

1. Stephanie May, unpublished memoirs, 224. Stephanie never completed her memoirs, and they currently consist of several discrete

bundles of double-spaced typed pages that are not always consecutively numbered. In referencing quotations from the memoirs in these notes, I've used her page numbers when possible to indicate the quotations' sequence in her larger narrative.

2. Whether the human species will survive the problems it creates for itself in the future, though, is a different matter. Some smart futurists and scientists argue cogently that rogue super-intelligent machines or lethal genetically engineered diseases could wipe us out entirely. See, for instance, Nick Bostrom, "The Vulnerable World Hypothesis," *Global Policy* 10, no. 4 (September 2019).
3. Basic introductions to complexity science are Melanie Mitchell, *Complexity: A Guided Tour* (Oxford: Oxford University Press, 2011), and Murray Gell-Mann, "What Is Complexity? Remarks on Simplicity and Complexity by the Nobel Prize–Winning Author of *The Quark and the Jaguar*," *Complexity* 1, no. 1 (September/October 1995), doi: 10.1002/cplx.6130010105.
4. Emergence is a philosophical minefield. Generally, scholars distinguish between "epistemological emergence," where a system's properties appear novel because we have inadequate knowledge of the system, so we can't fully predict its behavior, and "ontological emergence," where a system's properties are truly novel—that is, not reducible, even with complete knowledge, to the system's underlying parts and their interactions. For a useful general framework, see J. de Haan, "How Emergence Arises," *Ecological Complexity* 3 (2006): 293–301, doi: 10.1016/j.ecocom.2007.02.003.
5. Uncertainty can be of two kinds: "epistemic," which is a function of the observer's lack of knowledge of the world ("We don't understand the workings of the world enough, so we're uncertain about the probability of outcome X"), and "aleatory," which is a function of the universe's fundamental randomness ("Outcome X is intrinsically uncertain"). The uncertainty that arises from complexity is largely epistemic; however, both epistemic and aleatory uncertainty can provide a basis for hope.
6. Nathan Robinson makes a similar argument: "By taking a position that it is impossible to avoid calamity and extinction, a person asserts that they singlehandedly comprehend all possible futures." Nathan

J. Robinson, "Pessimism Is Suicide," *Current Affairs* (June 7, 2017).

7. Rebecca Solnit, *Hope in the Dark: Untold Histories, Wild Possibilities* (New York: Nation, 2004), 1.
8. Joseph Godfrey, *A Philosophy of Human Hope* (Dordrecht, Netherlands: Martinus Nijhoff, 1987), 29.
9. William Faulkner, "Banquet Speech," following receipt of the Nobel Prize in Literature, December 10, 1950, Stockholm, Sweden, https://www.nobelprize.org/prizes/literature/1949/faulkner/speech/.
10. Hesiod, *Works and Days*, trans. A.E. Stallings (New York: Penguin Classics, 2018), 6.
11. Philosophers and psychologists use the technical terms *intentional hope* to refer to hope that has an object (a state of mind "oriented to some desired state of affairs which is believed to be attainable," according to the *Routledge Encyclopedia of Philosophy*) and *dispositional hope* to refer to hope without an object ("a state of being hopeful"). Philip Stratton-Lake, "Hope," in *Routledge Encyclopedia of Philosophy*, general ed. Edward Craig, vol. 4 "Genealogy to Iqbal" (London: Routledge, 1998), 507.
12. Lear, *Radical Hope*, 103; Godfrey, *Human Hope*, 64.
13. C-SPAN, Senator Barack Obama 2004 Democratic National Convention Keynote Speech, July 27, 2004, Boston, MA, transcribed by author, retrieved from https://www.c-span.org/video/?182718-3/senator-barack-obama-2004-democratic-national-convention-keynote-speech.
14. Fred Bird, "Hope Makes a Difference" (unpublished manuscript, 2019), 36.
15. C-SPAN, Sarah Palin Remarks to Tea Party Convention, February 6, 2010, Nashville, TN, transcribed by author, retrieved from https://www.c-span.org/video/?291974-3/sarah-palin-remarks-tea-party-convention.
16. Stoics espouse clear-eyed realism, too. But they generally don't like hope, seeing it as a distraction from reality. While I agree with the Stoic position on realism, I'm convinced that hope has a key psychological function, so it's not something we should discard.
17. The distinction between "hope that" and "hope to" is examined by

R.S. Downie in "Hope," *Philosophy and Phenomenological Research* 24, no. 2 (December 1963): 248–51.

18. A person must perceive uncertainty to either "hope that" or "hope to," but the source of that uncertainty differs. With "hope that," the person is uncertain about the likelihood of the hoped-for future state; with "hope to," the person is mainly uncertain about his or her capacity to bring about that state.
19. Philip Pettit, "Hope and Its Place in Mind," *Annals AAPSS* 592 (March 2004): 161.
20. David Orr, *Hope Is an Imperative: The Essential David Orr* (Washington, DC: Island Press, 2010), 330.
21. Antonio Gramsci, "Notebook I (1929–1930)," *Prison Notebooks*, vol. I, ed. Joseph Buttigieg (New York: Columbia University Press, 1975), 172.

## CHAPTER 6: IMAGINE POSSIBILITY

1. Thomas Aquinas, *Summa Theologica*, First Part of the Second Part: "Question 40: Of the Irascible Passions, and First, of Hope and Despair"; "Article 5: Whether experience is a cause of hope?"; "Objection 3," http://www.documentacatholicaomnia.eu/03d/1225-1274,_Thomas_Aquinas,_Summa_Theologiae_%5B1%5D,_EN.pdf.
2. James R. Averill, George Catlin, and Kyum Koo Chon, *Rules of Hope* (New York: Springer-Verlag, 1990), 13, 16, and 33.
3. Stuart A. Kauffman, *Reinventing the Sacred: A New View of Science, Reason, and Religion* (New York: Basic Books, 2008), 64.
4. Michael Corballis, *The Recursive Mind: The Origins of Human Language, Thought, and Civilization* (Princeton, NJ: Princeton University Press, 2011), 6.
5. François Jacob, *The Possible and the Actual* (New York: Pantheon, 1982), 67.
6. Hope always involves an implicit judgment that the probability of the desired future state is neither zero nor 100 percent. This implicit judgment distinguishes hoping from wishing. A wish is a statement of desire—for a counterfactual state in the present or an imagined

future world—that has no probability assigned to it. One can and often does, for instance, wish for imagined worlds that are clearly impossible (as in "I wish I could fly like a bird").

7. May, unpublished memoirs, 219.
8. Philosophers and psychologists who study hope often use the unmodified term "uncertainty" somewhat imprecisely, as I do in this chapter, to mean a generalized lack of knowledge about present or future circumstances. But the twentieth-century American economist Frank Knight, in his 1921 book *Risk, Uncertainty, and Profit*, usefully distinguished risk from uncertainty. Risk, by his view, is a lack of knowledge of whether a specified outcome will occur—say, a string of ten straight heads in the toss of an unbiased coin—that can be reliably quantified using probability values derived from our past experience. In the case of an unbiased coin, we know from our experience with the behavior of such a coin that the probability of ten straight heads (or tails) is about a thousand to one. Uncertainty, in contrast, describes unquantifiable gaps in our knowledge, which are often the result of ignorance of the full range of possible outcomes. Perhaps, for instance, the coin can also land and stay on its edge, in addition to landing heads or tails, but when we toss it, we're not aware of that possible outcome. The term "deep uncertainty" highlights the profound nature of this ignorance.

   When our hope has an object, "uncertainty" usually means what Knight called risk: we say we're "uncertain" about whether the desired future will occur, because by our reckoning it's neither impossible nor inevitable; we believe its probability is greater than zero and less than 100 percent. But when we're searching for an object for our hope, our imagination can exploit deep uncertainty. It can identify previously unimagined outcomes—like that of the coin landing and staying on its edge—in the domain of the unknown. In doing so, our imagination pulls those potential outcomes into the domain of the known, where we can then begin to think about, and gather evidence on which to base, estimates of their probabilities.
9. Manuel DeLanda, *A New Philosophy of Society: Assemblage Theory and Social Complexity* (New York: Continuum, 2006), 1: "[Most] social entities, from small communities to large nation-states, would

disappear altogether if human minds ceased to exist. In this sense, social entities are not mind-independent."

10. An archetypal example is Newton's second law of motion, F = Ma. It states that the sum of the forces (F) acting on an object is equal to the mass (M) of that object times its acceleration (a). Time does play a role in this equation, because acceleration is a rate. (Specifically, acceleration is the second derivative of the position of an object with respect to time.) The law itself, though, and the mathematical "objects" that it defines—say, a parabola describing the motion of a falling body—exist outside time. The mathematical statement of this law is also an equation, and we can say that equations are instantaneously true, because their variables are mathematically linked to each other in the same instant. The positions of terms on the left and right of the equal sign don't indicate the progress of time; force, in other words, doesn't happen before mass or acceleration, just because one reads the *F* first from left to right.
11. Philosophers refer to the view that the past, present, and future happen simultaneously and are equally real (and that there's therefore no objective flow of time) as "eternalism." They usually argue it's a logical consequence of Einstein's conclusions in his special theory of relativity about what's called the "relativity of simultaneity." But the view is also common in Christian theology—in the idea that God transcends space and time and creates simultaneously the past, present, and future in the Divine Mind—and it can be traced to the ancient Greek philosopher Parmenides. Its most prominent modern proponent was the early-twentieth-century philosopher J.M.E. McTaggart.
12. Lee Smolin, *Time Reborn: From the Crisis of Physics to the Future of the Universe* (Boston: Houghton Mifflin Harcourt, 2013), 257.
13. All the laws and constants that scientists believe anchor reality probably evolve over time, according to the Smolin-Unger thesis, although those relating to nature's basic forces like electromagnetism and gravity may change exceedingly rarely.
14. Reinhart Koselleck, *Futures Past: On the Semantics of Historical Time*, trans. Keith Tribe (New York: Columbia University Press, 2004); for an interpretation of Koselleck's view of time and its relevance

to hope, see Peter Burke, "Does Hope Have a History?" *Estudos Avançados* 26 (2012).

15. Karl Popper, *Unended Quest: An Intellectual Biography* (London: Taylor & Francis, 2005), 149.
16. Isaiah Berlin, "Historical Inevitability, August Comte Memorial Trust Lecture, No.1, Delivered on 12 May 1953 at the London School of Economics and Political Science"(Oxford: Oxford University Press, 1957).
17. Smolin, *Time Reborn*, 258.
18. Emily Atkin, "The Power and Peril of 'Climate Disaster Porn,'" *The New Republic* (July 10, 2017).
19. If done precisely, the exercise would involve estimating a probability for each discrete amount of difference one could make in reducing the problem's severity (from no difference to solving the problem entirely) and then extracting from this "probability distribution" the most likely amount of difference.

## CHAPTER 7: COURAGE BEYOND THE EDGE

1. Patrick Henry, "Give Me Liberty or Give Me Death," March 23, 1775, text available at The Avalon Project, Yale Law School, https://avalon.law.yale.edu/18th_century/patrick.asp. Although Henry's speech is among the most famous in American history, he spoke without notes, and no exact record of his words exists.
2. Quoted in Studs Terkel, *Hope Dies Last: Keeping the Faith in Troubled Times* (New York: New Press, 2003), 42.
3. Baruch Spinoza, "Proposition 47: The Emotions of Hope and Fear Cannot be Good in Themselves, Scholium," in *Ethics: Treatise on the Emendation of the Intellect and Selected Letters,* trans. Samuel Shirley, ed. Seymour Feldman (Indianapolis: Hackett, 1992), 181.
4. Derrick Jensen, "Beyond Hope," *Orion Magazine*, May 2, 2006, https://orionmagazine.org/article/beyond-hope/.
5. Seneca, *Letters from a Stoic, Epistulae Morales ad Lucilium*, trans. Robin Campbell (New York: Penguin, 1969), 38.
6. "Carnival of the Senses: A Conversation with Michael Taussig," in Mary Zournazi, *Hope: New Philosophies of Change* (London: Psychology Press, 2003), 45.

7. Jensen, "Beyond Hope."
8. Paul Kingsnorth, "On the Correct Management of Despair," The Dark Mountain Project, August 31, 2011, https://dark-mountain.net/on-the-correct-management-of-despair/.
9. Aristotle, "Book II, Part 7," *Nicomachean Ethics*, trans. W.D. Ross, Internet Classics Archive, http://classics.mit.edu/Aristotle/nicomachaen.3.iii.html.
10. Ernst Bloch, "Introduction," *The Principle of Hope*, vol. 1, trans. Neville Plaice, Stephen Plaice, and Paul Knight (Cambridge, MA: MIT Press, 1986), 3.

## CHAPTER 8: THE FALSE PROMISE OF TECHNO-OPTIMISM

1. Jonathan L. Bamber, Michael Oppenheimer, Robert E. Kopp, Willy P. Aspinal, and Roger Cooke, "Ice Sheet Contributions to Future Sea-Level Rise from Structured Expert Judgment," *Proceedings of the National Academy of Sciences* 116, no. 23 (June 4, 2019): 11195–200; Alexander Nauels, Joeri Rogelj, Carl-Friedrich Schleussner, Malte Meinshausen, and Matthias Mengel, "Linking Sea Level Rise and Socioeconomic Indicators under the Shared Socioeconomic Pathways," *Environmental Research Letters* 12 (2017), doi: 10.1088/1748-9326/aa92b6; Dewi Le Bars, Sybren Drijfhout, and Hylke de Vries, "A High-End Sea Level Rise Probabilistic Projection Including Rapid Antarctic Ice Sheet Mass Loss," *Environmental Research Letters* 12 (2017), doi: 10.1088/1748-9326/aa6512.
2. Victor J. Polyak, Bogdan P. Onac, Joan J. Fornós, Carling Hay, Yemane Asmerom, Jeffrey A. Dorale, Joaquín Ginés, Paola Tuccimei, and Angel Ginés, "A Highly Resolved Record of Relative Sea Level in the Western Mediterranean Sea during the Last Interglacial Period," *Nature Geoscience* 11 (2018): 860–64, doi: 10.1038/s41561-018-0222-5.
3. Morgan E. Eisenlord, Maya L. Groner, Reyn M. Yoshioka, Joel Elliott, Jeffrey Maynard, Steven Fradkin, Margaret Turner, Katie Pyne, Natalie Rivlin, Ruben van Hooidonk, and C. Drew Harvell, "Ochre Star Mortality during the 2014 Wasting Disease Epizootic: Role of Population Size Structure and Temperature," *Philosophical*

*Transactions of the Royal Society B* 371 (January 25, 2017), doi: 10.1098/rstb.2015.0212.

4. In chapter 6, I proposed that objects of our hope must fall in the zone between the impossible and the inevitable. Outcomes that we perceive as impossible are strong structural *constraints* on our future possibilities and our honest hope, while the mental space between the impossible and inevitable creates structural *opportunities* to imagine objects for that hope.
5. See chapter 10, "Techno-Hubris," of *The Ingenuity Gap* for an example of this kind of analysis.
6. Robert J. Gordon, "Is U.S. Economic Growth Over? Faltering Innovation Confronts the Six Headwinds," Working Paper 18315 (Cambridge, MA: National Bureau of Economic Research, 2012), 3–7, http://www.nber.org/papers/w18315.
7. Erik Brynjolfsson and Andrew McAfee, *The Second Machine Age: Work, Progress, and Prosperity in a Time of Brilliant Technologies* (New York: Norton, 2014); Ruth DeFries, *The Big Ratchet: How Humanity Thrives in the Face of Natural Crisis* (New York: Basic Books, 2014); Peter H. Diamandis and Steven Kotler, *Abundance: The Future Is Better Than You Think* (New York: Free Press, 2012); Gregg Easterbrook, *It's Better Than It Looks: Reasons for Optimism in an Age of Fear* (New York: Public Affairs, 2018); Ray Kurzweil, *The Singularity Is Near: When Humans Transcend Biology* (New York: Viking, 2005); Matt Ridley, *The Rational Optimist: How Prosperity Evolves* (New York: HarperCollins, 2010); and Steven Pinker, *Enlightenment Now: The Case for Reason, Science, Humanism, and Progress* (New York: Penguin, 2018).
8. Angus Deaton, "Getting Better All the Time," *New York Times Book Review*, March 2, 2018. Deaton writes: "Much of what [Easterbrook] says is right, but much is not, or is wishful thinking, or sounds wildly optimistic, but does not seem to be documented and so is uncheckable. In the end, he weakens his case."
9. Diamandis and Kotler, *Abundance*, 9.
10. World Bank, *Piecing together the Poverty Puzzle* (Washington, DC: World Bank, 2018), 1, https://openknowledge.worldbank.org/bitstream/handle/10986/30418/9781464813306.pdf.

11. Will Koehrsen, "Has Global Violence Declined? A Look at the Data," Towards Data Science, January 5, 2019, https://towardsdatascience.com/has-global-violence-declined-a-look-at-the-data-5af708f47fba.
12. Anne Buchanan and Kenneth Weiss, "Things Genes Can't Do: Simplistic Ideas of How Genes 'Cause' Traits Are No Longer Viable: Life Is an Orderly Collection of Uncertainties," Aeon, April 9, 2013, https://aeon.co/essays/dna-is-the-ruling-metaphor-of-our-age.

### CHAPTER 9: THE WORLD TO COME TODAY

1. A.H. Maslow, "A Theory of Human Motivation," *Psychological Review* 50 (1943): 370–96.
2. Louis Tay and Ed Diener, "Needs and Subjective Well-Being around the World," *Journal of Personality and Social Psychology* 101, no. 2 (2011): 354–65, doi: 10.1037/a0023779.
3. N.J. Hagens, "Economics for the Future: Beyond the Superorganism," *Ecological Economics* 169 (2020), doi: 10.1016/j.ecolecon.2019.106520.
4. I develop this argument fully in *The Upside of Down*.
5. The most significant recent reports on Earth's environmental crisis are: Intergovernmental Science Policy Platform on Biodiversity and Ecosystem Services (IPBES), *Summary for Policymakers of the Global Assessment Report on Biodiversity and Ecosystem Services of the Intergovernmental Science-Policy Platform on Biodiversity and Ecosystem Services*, ed. S. Díaz, J. Settele, E.S. Brondizio, H.T. Ngo, M. Guèze, J. Agard, A. Arneth, P. Balvanera, K.A. Brauman, S.H.M. Butchart, K.M.A. Chan, L.A. Garibaldi, K. Ichii, J. Liu, S.M. Subramanian, G.F. Midgley, P. Miloslavich, Z. Molnár, D. Obura, A. Pfaff, S. Polasky, A. Purvis, J. Razzaque, B. Reyers, R. Roy Chowdhury, Y.J. Shin, I.J. Visseren-Hamakers, K.J. Willis, and C.N. Zayas (Bonn: IPBES secretariat, 2019); Intergovernmental Panel on Climate Change (IPCC), *Summary for Policymakers: Global Warming of 1.5°C: An IPCC Special Report on the Impacts of Global Warming of 1.5°C above Pre-industrial Levels and Related Global Greenhouse Gas Emission Pathways, in the Context of Strengthening the Global Response to the Threat of Climate Change, Sustainable Development, and Efforts to*

*Eradicate Poverty*, ed. V. Masson-Delmotte, P. Zhai, H.O. Pörtner, D. Roberts, J. Skea, P.R. Shukla, A. Pirani, W. Moufouma-Okia, C. Péan, R. Pidcock, S. Connors, J.B.R. Matthews, Y. Chen, X. Zhou, M.I. Gomis, E. Lonnoy, T. Maycock, M. Tignor, and T. Waterfield (Geneva: World Meteorological Organization, 2018); IPCC, *Summary for Policymakers: IPCC Special Report on the Ocean and Cryosphere in a Changing Climate*, ed. H.O. Pörtner, D.C. Roberts, V. Masson-Delmotte, P. Zhai, M. Tignor, E. Poloczanska, K. Mintenbeck, A. Alegría, M. Nicolai, A. Okem, J. Petzold, B. Rama, and N. Weyer (Geneva: World Meteorological Organization, 2019); and IPCC, *Summary for Policymakers (approved draft): IPCC Special Report Climate Change, Desertification, Land Degradation, Sustainable Land Management, Food Security, and Greenhouse Gas Fluxes in Terrestrial Ecosystems*, 2019, https://www.ipcc.ch/site/assets/uploads/2019/08/Edited-SPM_Approved_Microsite_FINAL.pdf.

6. The statistics in this and the following paragraph are drawn from United Nations, Department of Economic and Social Affairs, Population Division, *World Population Prospects 2019: Highlights* (New York: United Nations, 2019), https://population.un.org/wpp/Publications/Files/WPP2019_Highlights.pdf.
7. See, for instance, Darrell Bricker and John Ibbitson, *Empty Planet: The Shock of Global Population Decline* (New York: Crown, 2019).
8. Data on per capita carbon emissions are generated by the Carbon Dioxide Information Analysis Center, Environmental Sciences Division, Oak Ridges National Laboratory, and published by the World Bank at https://data.worldbank.org/indicator/EN.ATM.CO2E.PC.
9. Scientists estimate that Earth's average surface temperature in 2019 is around 14.5 degrees Celsius or 58.1 degrees Fahrenheit.
10. Thorsten Mauritsen and Robert Pincus, "Committed Warming Inferred from Observations," *Nature Climate Change* 7 (2017): 652–55, doi: 10.1038/nclimate3357.
11. Adrian Raftery, Alec Zimmer, Dargan M.W. Frierson, Richard Startz, and Peiran Liu, "Less than 2°C Warming by 2100 Unlikely," *Nature Climate Change* 7 (2017): 637–41, doi: 10.1038/nclimate3352.
12. Jeremy S. Pal and Elfatih A.B. Eltahir, "Future Temperature in

Southwest Asia Projected to Exceed a Threshold for Human Adaptability," *Nature Climate Change* 6 (2016): 197–200, doi: 10.1038/nclimate2833; and Im Eun-Soon, Jeremy S. Pal, and Elfatih A.B. Eltahir, "Deadly Heat Waves Projected in the Densely Populated Agricultural Regions of South Asia," *Science Advances* 3, no. 8 (August 2, 2017), doi: 10.1126/sciadv.1603322.

13. IPCC, *Summary for Policymakers (approved draft): Special Report Climate Change, Desertification, Land Degradation,* 16.
14. Philip K. Thornton, "Recalibrating Food Production in the Developing World: Global Warming Will Change More Than Just the Climate," *CCAFS Policy Brief no. 6,* CGIAR Research Program on Climate Change, Agriculture and Food Security (CCAFS), 2012, https://cgspace.cgiar.org/handle/10568/24696.
15. For a comprehensive overview of threats to global biodiversity, see World Wildlife Fund, *Living Planet Report 2018: Aiming Higher,* ed. M. Grooten and R.E.A. Almond (Gland, Switzerland: WWF, 2018).
16. Gerardo Ceballos, Paul R. Ehrlich, and Rodolfo Dirzo, "Biological Annihilation Via the Ongoing Sixth Mass Extinction Signaled by Vertebrate Population Losses and Declines," *Proceedings of the National Academy of Sciences,* 114, no. 30 (July 25, 2017), doi: 10.1073/pnas.1704949114.
17. Stuart L. Pimm, Gareth J. Russell, John L. Gittleman, Thomas M. Brooks, "The Future of Biodiversity," *Science* 269, no. 5222 (July 21, 1995): 347–50, doi: 10.1126/science.269.5222.347.
18. World Wildlife Fund, *Living Planet Report 2018,* 7: "The Living Planet Index . . . tracks the state of global biodiversity by measuring the population abundance of thousands of vertebrate species around the world. The latest index shows an overall decline of 60 percent in population sizes between 1970 and 2014." Elizabeth Pennisi, "Billions of North American Birds Have Vanished," *Science* 365, no. 6459 (September 20, 2019): 1228–29, doi: 10.1126/science.365.6459.1228.
19. See chapter 2 in *The Upside of Down* for a non-technical overview of the implications of the changing energy cost of energy. For a technical analysis, see Adam R. Brandt, "How Does Energy Resource Depletion Affect Prosperity? Mathematics of a Minimum Energy

Return on Investment (EROI)," *Biophysical Economics and Resource Quality* 2, article 2 (2017), doi: 10.1007/s41247-017-0019-y.

20. Humankind's responses to declining resource quality imply another thing, too. Should our species' high-technology civilization fail, neither human beings nor any other species will have another chance to build such a civilization on Earth. As the renowned English astronomer Fred Hoyle pointed out in the 1960s: "It has often been said that, if the human species fails to make a go of it here on Earth, some other species will take over the running. In the sense of developing intelligence this is not correct. We have, or soon will have, exhausted the necessary physical prerequisites so far as this planet is concerned. With coal gone, oil gone, high-grade metallic ores gone, no species however competent can make the long climb from primitive conditions to high-level technology. This is a one-shot affair. If we fail, this planetary system fails so far as intelligence is concerned." Or as the Canadian author Ronald Wright has said: "For all its cruelties, civilization is precious, an experiment worth continuing. It is also precarious: as we climbed the ladder of progress, we kicked out the rungs below." Fred Hoyle, *Of Men and Galaxies* (Seattle: University of Washington Press, 1964); and Ronald Wright, *A Short History of Progress* (Toronto: House of Anansi Press, 2004), 34.
21. Oxfam, *Reward Work, Not Wealth: To End the Inequality Crisis, We Must Build an Economy for Ordinary Working People, Not for the Rich and Powerful*, Oxfam Briefing Paper (January 2018), 10, https://oi-files-d8-prod.s3.eu-west-2.amazonaws.com/s3fs-public/file_attachments/bp-reward-work-not-wealth-220118-en.pdf.
22. Federal Reserve Bank of St. Louis, "Employed Full Time: Median Usual Weekly Real Earnings: Wage and Salary Workers: 16 Years and Over: Men," https://fred.stlouisfed.org/series/LES1252881900Q.
23. Lawrence Mishel and Jessica Schieder, "CEO Pay Remains High Relative to the Pay of Typical Workers," Table 1: CEO Compensation, CEO-to Worker Compensation Ratio, and Stock Prices (2016 dollars), 1965–2016, Economic Policy Institute, July 20, 2017, https://www.epi.org/publication/ceo-pay-remains-high-relative-to-the-pay-of-typical-workers-and-high-wage-earners/.

24. Branko Milanovic, *Capitalism Alone: The Future of the System That Rules the World* (Cambridge, MA: Belknap, 2019), especially 44 and 101. Milanovic shows that disposable income inequality in the United States rose from a Gini coefficient of 0.35 in 1974 to 0.41 in 2016, while overall rural and urban income inequality in China rose from a Gini coefficient of 0.3 in 1985 to about 0.47 in 2015. For further evidence on rising inequality in the United States, Europe, and China since the 1980s, see Gabriel Zucman, "Global Wealth Inequality," Working Paper 25462 (Washington, DC: National Bureau of Economic Research, 2019), http://www.nber.org/papers/w25462.
25. United Nations High Commissioner for Refugees, "Figures at a Glance: Statistical Yearbooks," https://www.unhcr.org/en-us/figures-at-a-glance.html.
26. A particularly compelling exercise in forecasting through fictional narrative is the novella by the historians of science Naomi Oreskes and Erik M. Conway titled "The Collapse of Western Civilization: A View from the Future," first published in *Dædalus* 142, no. 1 (Winter 2013), 40–58. The authors tell the story of a global climate catastrophe through the eyes of a historian in the twenty-fourth century. "The most astounding fact," the historian writes, "is that the victims *knew what was happening and why*. Indeed, they chronicled it in detail precisely *because* they knew that fossil fuel combustion was to blame. Historical analysis also shows that Western civilization had the technological know-how and capability to effect an orderly transition to renewable energy, yet the available technologies were not implemented in time."
27. Oran R. Young, Frans Berkhout, Gilberto C. Gallopin, Marco A. Janssen, Elinor Ostrom, and Sander van der Leeuw, "The Globalization of Socio-Ecological Systems: An Agenda for Scientific Research," *Global Environmental Change* 16 (2006): 304–16, doi: 10.1016/j.gloenvcha.2006.03.004.

## CHAPTER 10: A CONTEST OF WITS

1. Complex adaptive systems often incorporate "agents"—independent, complex subsystems—each of which has an internal model of

its external environment that guides the agent's responses to changes in that environment. Complexity scientists often refer to these internal models as "schemas." In practice, all known complex adaptive systems involve life in some way; the agents are either living entities themselves or they're dependent on living entities. A mammalian gut—say, in a rabbit, a cow, or a human being—is a complex adaptive system, and an individual *E. coli* bacterium in that gut is an agent; the bacterium's genetic code, which guides its response to its environment, is its schema. As the larger organism's diet changes, some strains of bacteria proliferate and others dwindle in numbers; in the process, the gut adapts to the change or "learns." A society is a complex adaptive system. Its agents include corporations, community groups, government bureaucracies, and, of course, individual people. The schemas for these agents are both biological (for instance, the genetic codes of the people involved) and, importantly, symbolic, in the form of information (beliefs and values) encoded in people's cerebral cortexes. For complexity scientists, a worldview is a key part of a person's symbolic schema, because it helps guide that person's response to his or her social environment. For an account of complex adaptive systems and schemas, see Murray Gell-Mann, "The Simple and the Complex" in *Complexity, Global Politics, and National Security*, ed. D. Alberts and T. Czerwinski (Washington, DC: National Defense University, 1997), 2–12.

2. Mrs. Franklin D. Roosevelt (signed "Eleanor Roosevelt"), letter to Stephanie May, December 5, 1956.
3. Elizabeth Ives, letter to Stephanie May, November 17, 1956.
4. Gerald W. Johnson, letter to Stephanie May, November 23, 1956.
5. Rachael Beddoe, Robert Costanza, Joshua Farley, Eric Garza, Jennifer Kent, Ida Kubiszewski, Luz Martinez, Tracy McCowen, Kathleen Murphy, Norman Myers, Zach Ogden, Kevin Stapleton, and John Woodward, "Overcoming Systemic Roadblocks to Sustainability: The Evolutionary Redesign of Worldviews, Institutions, and Technologies," *Proceedings of the National Academy of Sciences* 106, no. 8 (February 24, 2009): 2483–89, doi: 10.1073/pnas.0812570106. Earlier development of similar ideas can be found in Gerhard

Lenski, *Ecological-Evolutionary Theory: Principles and Applications* (Boulder, CO: Paradigm, 2005); and Robert Cox, "Social Forces, States, and World Orders: Beyond International Relations Theory," in *Approaches to World Order*, ed. Timothy J. Sinclair (Cambridge, UK: Cambridge University Press, 1996 [1981]), 85–123, especially 98, where Cox analyzes the interactions within social structures of "ideas," "material capabilities," and "institutions."

6. Annick de Witt provides a comprehensive discussion of theory and research on worldviews in *Worldviews and the Transformation to Sustainable Societies: An Exploration of the Cultural and Psychological Dimensions of Our Global Environmental Challenges* (PhD dissertation, Utrecht University, 2013), doi: 10.13140/RG.2.1.4492.8406. See also Francis Fukuyama, *The Origins of Political Order: From Prehuman Times to the French Revolution* (New York: Farrar, Straus, and Giroux, 2011), 442: "People in all human societies create mental models of reality. These mental models attribute causality to various factors—oftentimes invisible ones—and their function is to make the world more legible, predictable, and easy to manipulate."
7. The classic statement of this idea of institutions is Douglass C. North, *Institutions, Institutional Change, and Economic Performance* (New York: Cambridge University Press, 1990). North writes (p. 3) that "institutions are the rules of the game in a society, or more formally, are the humanly devised constraints that shape human interaction."
8. This understanding of technology is elaborated in W. Brian Arthur, *The Nature of Technology: What It Is and How It Evolves* (New York: Free Press, 2011).
9. Karl Marx similar distinguished between a society's economic base of material conditions (roughly corresponding to a WIT's "technologies"), its structure ("institutions"), and its superstructure ("worldviews"). However, Marx organized these categories hierarchically, arguing that the superstructure (of ideas, ideologies, and worldviews) at the top of the hierarchy was epiphenomenal, being completely derivative of society's underlying, foundational material conditions and relationships. Beddoe and her coauthors, along with scholars such as Lenski and Cox (previously cited), arrange the triad horizontally rather than vertically. This shift from Marx's vertical to

a horizontal (or circular) relationship gives ideas, ideologies, and worldviews causal efficacy. Much of twentieth-century Marxist thought, including that of Karl Mannheim, Antonio Gramsci, and Louis Althusser, was an attempt to correct the obviously flawed assumption in Marx's original formulation. See chapter 2, "Overcoming Illusions: How Ideologies Came to Stay," in Michael Freeden, *Ideology: A Very Short Introduction* (Oxford: Oxford University Press), 12–30, for a summary. The causal "equality" of each element is a key reason the WIT formulation is so useful.

10. Key discussions include: Michael Foucault, *The History of Sexuality, Volume I: An Introduction,* trans. Robert Hurley (New York: Vintage, 1990); Michael Barnett and Raymond Duvall, *Power in Global Governance* (Cambridge, UK: Cambridge University Press, 2005); Steven Lukes, *Power: A Radical View,* 2nd ed. (Houndmills, Basingstoke, Hampshire, UK: Palgrave Macmillan, 2005); and Norbert Elias, "Power and Civilization," *Journal of Power* 1, no. 2 (2008): 135–42, doi: 10.1080/17540290802309540.
11. Schubert and Lapp, "Radiation Dangers," 10.

## CHAPTER 11: WHY IS POSITIVE CHANGE SO HARD?

1. Kathryn Stevenson and Nils Peterson, "Motivating Action through Fostering Climate Change Hope and Concern and Avoiding Despair among Adolescents," *Sustainability* 8, no. 6 (2016), doi: 10.3390/su8010006. Drawing on his research on the psychological effects of the fear of nuclear war, famed US psychiatrist Robert Jay Lifton discusses the "psychic numbing" that the threat of climate change can induce in people. Lifton, *The Climate Swerve: Reflections on Mind, Hope, and Survival* (New York: New Press, 2017).
2. For a general introduction, see John Cassidy, *How Markets Fail: The Logic of Economic Calamites* (New York: Farrar, Straus and Giroux, 2009).
3. Joseph E. Stiglitz, "Government Failure vs. Market Failure," in *Governments and Markets: Toward a New Theory of Regulation,* eds. Edward J. Balleisen and David A. Moss (Cambridge, UK: Cambridge University Press, 2010), 13–51.
4. The late American economist Mancur Olson provided perhaps the

most persuasive analysis of the underlying economic incentives that encourage the entrenchment of vested interests, or what he called "narrow distributional coalitions," in a society. See Olson, *The Logic of Collective Action: Public Goods and the Theory of Groups* (Cambridge, MA: Harvard University Press, 1965 and 1971); and Olson, *The Rise and Decline of Nations: Economic Growth, Stagflation, and Social Rigidities* (New Haven, CT: Yale University Press, 1982).

5. Dwight D. Eisenhower, "Transcript of President Dwight D. Eisenhower's Farewell Address (1961)," https://www.ourdocuments.gov/doc.php?flash=false&doc=90&page=transcript.
6. Financial Stability Board, "2018 List of Global Systemically Important Banks (G-SIBs)," https://www.fsb.org/wp-content/uploads/P161118-1.pdf.
7. On vested interest and the decline of nations, see Olson, *Logic* and *Decline*. On social rigidities, complexity and the collapse of empires, see Joseph A. Tainter, *The Collapse of Complex Societies* (Cambridge, UK: Cambridge University Press, 1988).
8. Matthias Schmelzer, *The Hegemony of Growth: The OECD and the Making of the Economic Growth Paradigm* (Cambridge, UK: Cambridge University Press, 2016). For a historical treatment of the origins of the growth paradigm, see Stephen J. Purdey, *Economic Growth, the Environment and International Relations* (Oxford: Routledge, 2010).
9. Eric D. Beinhocker, *The Origin of Wealth: Evolution, Complexity, and the Radical Remaking of Economics* (Boston: Harvard Business Press, 2006).
10. United Nations World Commission on Environment and Development, "Chapter 1. A Threatened Future. Section 49" in *Report of the World Commission on Environment and Development: Our Common Future* (1987), http://www.un-documents.net/ocf-01.htm.
11. Kuznets didn't propose this specific relationship. Rather, he posited an analogous relationship between a society's per capita income and its level of economic inequality, in which rising income is at first associated with an increase, and then subsequently with a decrease, in inequality. On the environmental Kuznets curve (EKC), see Dimitri Kaika and Efthimios Zervas, "The Environmental

Kuznets Curve (EKC) Theory—Part A: Concept, Causes and the $CO_2$ Emissions Case," *Energy Policy* 62 (2013): 1392–1402, doi: 10.1016/j.enpol.2013.07.131; and Kaika and Zervas, "The Environmental Kuznets Curve (EKC) Theory. Part B: Critical Issues," *Energy Policy* 62 (2013): 1403–11, doi: 10.1016/j.enpol.2013.07.130.

12. Figures on changes in national per capita $CO_2$ emissions derived from the Global Carbon Atlas, "$CO_2$ Emissions. Territorial Per Capita," http://www.globalcarbonatlas.org/en/CO2-emissions.
13. The statistics in this and the next paragraph are from Steven Nadel, Neal Elliott, and Therese Langer, *Energy Efficiency in the United States: 35 Years and Counting,* Report E1502 (Washington, DC: American Council for an Energy Efficient Economy, 2015), iv, https://www.aceee.org/sites/default/files/publications/researchreports/e1502.pdf.
14. For additional evidence that economic growth cannot be "decoupled" from environmental damage, see James D. Ward, Paul C. Sutton, Adrian D. Werner, Robert Costanza, Steve H. Mohr, and Craig T. Simmons, "Is Decoupling GDP Growth from Environmental Impact Possible?", *PLOS One* (October 14, 2016), doi: 10.1371/journal.pone.0164733.
15. Specifically, the argument assumes that economies can easily substitute human-made capital for natural capital or, in technical terms, that the elasticity of substitution of the former for the latter is near-infinite.
16. I first articulated these equivalencies in a speech in Essen, Germany, on June 8, 2009, "The Great Transformation: Climate Change as Cultural Change," available at https://homerdixon.com/the-great-transformation-climate-change-as-cultural-change/.
17. World Bank, *Piecing together the Poverty Puzzle*, 1 and 6.
18. Benjamin M. Friedman, *The Moral Consequences of Economic Growth* (New York: Knopf, 2005).
19. See chapter 8, "No Equilibrium," in *The Upside of Down* for a full account. Many historical and contemporary thinkers, including Keynes and Marx, have made similar arguments.
20. I believe a fourth, but somewhat subordinate, equivalency also often operates in people's minds: growth equals solvency. Countries,

companies, and households rely on growth to maintain their solvency over time: if incomes and wealth rise, debt becomes a smaller proportion of that wealth, and payments on that debt a smaller fraction of income.

21. Robert J. Gordon, "The Demise of U.S. Economic Growth: Restatement, Rebuttal, and Reflections," Working Paper 19895 (Cambridge, MA: National Bureau of Economic Research, 2014, http://www.nber.org/papers/w19895; see also Gordon, *The Rise and Fall of American Growth: The U.S. Standard of Living since the Civil War* (Princeton, NJ: Princeton University Press, 2016).
22. Burke and Tanutama, "Climatic Constraints."

## CHAPTER 12: SHOCK CASCADES

1. The arguments in this and the next paragraphs are elaborated in Young et al., "Globalization of Socio-Ecological Systems," 308–11.
2. Fragkiskos Papadopoulus, Maksim Kitsak, M. Ángeles Serrano, Marián Boguña, and Dimitri Krioukov, "Popularity Versus Similarity in Growing Networks," *Nature* 489 (September 27, 2012): 537–40.
3. The self-reinforcing (positive) feedback described in this paragraph, where the incentive for an agent to join a network increases as the network adds more members, produces what economists call "increasing returns to scale" and can contribute to the "lock-in" of sub-optimal technologies, institutions, and social relations. The complexity economist W. Brian Arthur pioneered research on this topic. See Arthur, "Self-Reinforcing Mechanisms in Economics," in *The Economy as an Evolving Complex System,* Proceedings of a workshop sponsored by the Santa Fe Institute (Boca Raton, FL: Taylor & Francis, 1988).
4. Marten Scheffer, Stephen R. Carpenter, Timothy M. Lenton, Jordi Bascompte, William Brock, Vasilis Dakos, Johan van de Koppel, Ingrid A. van de Leemput, Simon A. Levin, Egbert H. van Nes, Mercedes Pascual, and John Vandermeer, "Anticipating Critical Transitions," *Science* 338 (October 19, 2012): 344–48, doi: 10.1126/science.1225244.
5. Dirk Helbing, "Globally Networked Risks and How to Respond," *Nature* 497 (May 2, 2013): 51–59, doi: 10.1038/nature12047.

6. Historical global GDP growth figures calculated using World Bank and OECD national accounts data, "GDP growth (annual %)," https://data.worldbank.org/indicator/NY.GDP.MKTP.KD.ZG?end=2018&start=1961.
7. For a more detailed discussion, see Thomas Homer-Dixon, Brian Walker, Reinette Biggs, Anne-Sophie Crépin, Carl Folke, Eric F. Lambin, Garry Peterson, Johan Rockström, Marten Scheffer, Will Steffen, and Max Troell, "Synchronous Failure: The Emerging Causal Architecture of Global Crisis," *Ecology and Society* 20, no. 3 (2015).
8. Jichang Zhao, Daqing Li, Hillel Sanhedrai, Reuven Cohen, and Shlomo Havlin, "Spatio-Temporal Propagation of Cascading Overload Failures in Spatially Embedded Networks," *Nature Communications* 7 (2016), doi: 10.1038/ncomms10094.

## CHAPTER 13: A MESSAGE FROM MIDDLE-EARTH

1. J.R.R. Tolkien, "The Council of Elrond," Book Two, Chapter II, *The Lord of the Rings, Part 1: The Fellowship of the Ring* (London: HarperCollins, 1994 [first published by George Allen & Unwin, 1955]), 350.
2. The debate between the Elf king, Athrabeth Finrod, and the woman, Andreth, is published in J.R.R. Tolkien, *Morgoth's Ring: The History of Middle-earth, Volume 10, The Later Silmarillion, Part One, The Legends of Aman*, ed. Christopher Tolkien (New York: HarperCollins, 1994), 301–65. For a discussion of the distinction between Amdir and Estel, see Elise Ringo, "Beyond Good and Evil: The Complex Moral System of Tolkien's Middle-earth," Tor.com, December 29, 2017, https://www.tor.com/2017/12/29/beyond-good-and-evil-the-complex-moral-system-of-tolkiens-middle-earth/.
3. Tolkien, *Lord of the Rings*, 353.
4. Stuart Kauffman, "Chapter 8: High-Country Adventures," in *At Home in the Universe: The Search for Laws of Self-Organization and Complexity* (Oxford: Oxford University Press, 1995), 149–89.
5. This divide parallels the one I described in chapter 6 between environmentalists who argue that the truth should be told about the gravity of our environmental crisis, even if it's dismal, and those

who argue that dismal truths simply scare people into denial and helplessness, so the seriousness of the crisis should be soft-pedaled.

6. Kelly introduced this aphorism in a poster for the first Earth Day, on April 22, 1970. A year later he incorporated it into the Pogo comic strip shown.
7. Muzafer Sherif, O.J.Harvey, B. Jack White, William R. Hood, and Carolyn W. Sherif, *The Robbers Cave Experiment: Intergroup Conflict and Cooperation* (Middletown, CT: Wesleyan University Press, 1988 [first published by the Institute of Group Relations, University of Oklahoma, 1961]). The late, renowned American social psychologist Roger Brown provides a thoughtful assessment of this experiment and the role of superordinate goals in conflict resolution in Brown, *Social Psychology,* 2nd ed. (New York: Free Press, 1986), 535–39 and 610–13.
8. Jacob Poushter and Christine Huang, "Climate Change Still Seen as the Top Global Threat, but Cyberattacks a Rising Concern," Pew Research Center: Global Attitudes and Trends, February 10, 2019, https://www.pewresearch.org/global/2019/02/10/climate-change-still-seen-as-the-top-global-threat-but-cyberattacks-a-rising-concern/.
9. Carbon dioxide as a gas in the atmosphere doesn't have weight, of course, but it still has mass that can be measured in tonnes. The estimate of required carbon dioxide removal to keep temperatures from rising above 1.5 degrees by 2100 is provided in James Hansen, Makiko Sato, Pushker Kharecha, Karina von Schuckmann, David J. Beerling, Junji Cao, Shaun Marcott, Valerie Masson-Delmotte, Michael J. Prather, Eelco J. Rohling, Jeremy Shakun, Pete Smith, Andrew Lacis, Gary Russell, and Reto Ruedy, "Young People's Burden: Requirement of Negative $CO_2$ Emissions," *Earth System Dynamics* 8 (2017): 577–616; see especially Figure 10 (b) "Atmospheric $CO_2$ including effect of $CO_2$ extraction that increases linearly after 2020," and discussion of this figure on page 591.
10. The details of the $CO_2$ volume calculation are as follows: One tonne of $CO_2$ has about 540 cubic meters of volume at sea level. A half trillion tonnes therefore has a volume of about $540 \times 0.5 \times 10^{12}$ cubic meters, or $2.7 \times 10^{14}$ cubic meters. Using the standard volume of sphere equation, the radius of a sphere of this volume is the cube

root of $(3 \times 2.7 \times 10^{14})/4\pi$, or the cube root of $0.645 \times 10^{14}$, which equals 40,104 meters (using the root calculator at https://www.calculator.net/root-calculator.html). The sphere is therefore 80 kilometers in diameter. The volume of the Grand Canyon, according to the United States National Park Service (https://www.nps.gov/grca/learn/management/statistics.htm) is 4.17 trillion cubic meters, or about 1/65 of volume of a half-trillion tonnes of $CO_2$.

11. Rachel Warren, "The Role of Interactions in a World Implementing Adaptation and Mitigations Solutions to Climate Change," *Philosophical Transactions of the Royal Society A* 369, no. 1934 (January 13, 2011): 234, doi: 10.1098/rsta.2010.0271.
12. United Nations Environment Programme, "Table ES.1: Total Global Greenhouse Gas Emissions in 2030 under Difference Scenarios (Median and Tenth to Ninetieth Percentile Range), Temperature Implications and Resulting Emissions Gap," in *Executive Summary, Emissions Gap Report 2018* (Nairobi: UNEP, 2018), 8, https://wedocs.unep.org/bitstream/handle/20.500.11822/26879/EGR2018_ESEN.pdf?sequence=10. See also Benjamin M. Sanderson, Brian C. O'Neill, and Claudia Tebaldi, "What Would It Take to Achieve the Paris Temperature Targets?", *Geophysical Research Letters* 43 (2016): 7133–42, doi: 10.1002/2016GL069563.
13. I've had this image in my mind for many years. In 2017, the director Darren Aronofsky used a similar metaphor in his controversial film *Mother*, which he acknowledged was an allegory about climate change.
14. As listed by the United Nations, the Sustainable Development Goals are: 1. No Poverty; 2. Zero Hunger; 3. Good Health and Well-being; 4. Quality Education; 5. Gender Equality; 6. Clean Water and Sanitation; 7. Affordable and Clean Energy; 8. Decent Work and Economic Growth; 9. Industry, Innovation, and Infrastructure; 10. Reduced Inequality; 11. Sustainable Cities and Communities; 12. Responsible Consumption and Production; 13. Climate Action; 14. Life below Water; 15. Life on Land; 16. Peace, Justice, and Strong Institutions; and 17. Partnerships for the Goals. See https://sustainabledevelopment.un.org/?menu=1300.

## CHAPTER 14: FROM GONDOR TO WASHINGTON, DC

1. Two pioneering efforts to formulate a complex-systems approach to economics are Eric D. Beinhocker, *The Origin of Wealth: Evolution, Complexity, and the Radical Remaking of Economics* (Boston: Harvard Business Press, 2006): and W. Brian Arthur, *Complexity and the Economy* (Oxford: Oxford University Press, 2015). Kate Raworth also rethinks conventional economics in *Doughnut Economics: 7 Ways to Think Like a 21st Century Economist* (White River Junction, VT: Chelsea Green, 2017).
2. An important attempt to describe the macro features of an alternative economic arrangement that could be compatible with Earth's deteriorating material conditions is Robert Costanza, Gar Alperovitz, Herman Daly, Joshua Farley, Carol Franco, Tim Jackson, Ida Kubiszewski, Juliet Schor, and Peter Victor, *Building a Sustainable and Desirable Economy-in-Society-in-Nature* (Canberra: Australian National University), 2013, https://www.jstor.org/stable/j.ctt5hgz53. A classic early essay is Kenneth E. Boulding, "The Economics of the Coming Spaceship Earth," in *Environmental Quality in a Growing Economy*, ed. H. Jarrett (Baltimore: Johns Hopkins University Press, 1966), 3–14, http://www.ub.edu/prometheus21/articulos/obsprometheus/BOULDING.pdf.
3. Donella Meadows, *Leverage Points: Places to Intervene in a System* (Hartland, VT: Sustainability Institute: 1999), 1, http://www.donellameadows.org/wp-content/userfiles/Leverage_Points.pdf.
4. Meadows, *Leverage Points,* 14.
5. Associated Press, "Expert Calls A-Fallout Threat to Unborn," *Hartford Times*. January 17, 1957.
6. Stephanie May, "Fair Exchange," *New York Post*, January 23, 1957.
7. Stephanie May, "Gambling with Humanity," *New York Post*, June 1, 1957.
8. Stephanie May, unpublished memoirs, 235.
9. Elsie Carper, "Housewives Petition Ike to Halt Nuclear Testing," *Washington Post and Times Herald*, June 26, 1957, B3.
10. Meadows, *Leverage Points,* 13.
11. Transcribed by author from the video at https://www.youtube.com/watch?v=VFkQSGyeCWg.

12. The American author and founder of the Liology Institute, Jeremy Lent, has written a superb analysis of the need for—and the potential mechanisms that could produce—a species-wide worldview transformation. See Lent, *The Patterning Instinct: A Cultural History of Humanity's Search for Meaning* (Amherst, NY: Prometheus, 2017).
13. Meadows, *Leverage Points,* 18.
14. In their seminal article on WITs, Rachael Beddoe, Robert Costanza, and their co-authors similarly suggest that dominant worldviews largely determine what's institutionally and technologically possible: "These unconscious assumptions about how the world works provide the boundary conditions within which institutions and technologies are designed to function." Beddoe et al., "Overcoming Systemic Roadblocks," 2484.
15. Nils Gilman, vice president of the Berggruen Institute, refers to the dominant worldview as the "official future." It is normally, he says, a "necessary form of delusion. It represents a kind of ideological glue that holds a collectivity together by defining a shared horizon of expectations. It makes social and political peace possible and creates a basis for collective action." See Gilman, "The Official Future Is Dead! Long Live the Official Future!", *The American Interest*, October 30, 2017, https://www.the-american-interest.com/2017/10/30/official-future-dead-long-live-official-future/.

## CHAPTER 15: INTO THE MIND

1. "Pessimism about Children's Future Is Widespread in Most Economies," Pew Research Center: Global Attitudes & Trends, January 2, 2019, https://www.pewresearch.org/global/2018/09/18/expectations-for-the-future/global_economy_charts-02/.
2. Ipsos, *Global @dvisor: Power to the People* (January 2017), 41, https://www.ipsos.com/sites/default/files/2017-01/Power_to_the_people_survey-01-2017.pdf.
3. Ipsos, *Global Trends: Fragmentation, Cohesion & Uncertainty* (2017), 56, https://www.ipsos.com/sites/default/files/2017-07/Ipsos%20Global%20Trends%202017%20report.pdf.
4. Ipsos MORI, "Only a Third of Generation Y Think Their Generation Will Have Better Quality of Life Than Their Parents,"

March 11, 2016, https://www.ipsos.com/ipsos-mori/en-uk/only-third-generation-y-think-their-generation-will-have-better-quality-life-their-parents.

5. Some commentators cite Denmark and Germany as countries where the power of vested interests was overcome to speed a transition to low-carbon economies. But a close look at these cases reveals a more complicated story. In Denmark, most decarbonization happened because of shifts in the global economy that forced the country to abandon key carbon-intensive industries. And while Germany heavily subsidized the early rollout of renewable energy, the country shielded its coal industry from change for decades.
6. Ingrid Creppell, "The Concept of Normative Threat," *International Theory* 3, no. 3 (2011): 450–87, doi: 10.1017/S1752971911000170.
7. Ipsos, *Global Trends*, 55; Ipsos, *Global @dvisor*, 41.
8. Ronald F. Inglehart and Pippa Norris, *Trump, Brexit, and the Rise of Populism: Economic Have-Nots and Cultural Backlash,* Faculty Research Working Paper 16-026 (Cambridge, MA: Harvard Kennedy School, August 2016).
9. Ipsos, *Global Trends*, 55.
10. David Brooks, "The End of the Two-Party System," *New York Times*, February 12, 2018.
11. I shared this belief at the time, writing in *The Upside of Down* about how massive "open-source problem solving" could address humanity's challenges.
12. Psychiatrist Norman Doidge and information technology entrepreneur Jim Balsillie discuss these issues in "Can We Ever Kick Our Smartphone Addiction?", *Globe and Mail*, February 18, 2018, https://www.theglobeandmail.com/opinion/can-we-ever-kick-our-smartphone-addiction-jim-balsillie-and-norman-doidgediscuss/article37976255/.
13. Michelle Goldberg, "The Darkness Where the Future Should Be," *New York Times*, January 24, 2020.
14. Edmund Burke, "Part II. Section II. Terror," *A Philosophical Inquiry into the Origin of Our Ideas of the Sublime and the Beautiful with an Introductory Discourse Concerning Taste, and Several Other Additions*

(1757), https://www.gutenberg.org/files/15043/15043-h/15043-h.htm#A_PHILOSOPHICAL_INQUIRY.

15. Anthony Leiserowitz, Edward Maibach, Seth Rosenthal, John Kotcher, Matthew Ballew, Matthew Goldberg, and Abel Gustafson, "1. Global Warming Beliefs," in *Climate Change in the American Mind: December 2018* (Yale Program on Climate Change Communication, January 22, 2019), https://climatecommunication.yale.edu/publications/climate-change-in-the-american-mind-december-2018/3/.
16. The integrity of people we might call "professional contrarians," such as lobbyists funded by and promoting the interests of fossil-fuel companies, is another matter.
17. Two noteworthy treatments are George Marshall, *Don't Even Think About It: Why Our Brains Are Wired to Ignore Climate Change* (New York: Bloomsbury, 2014); and Kari Marie Norgaard, *Living in Denial: Climate Change, Emotions, and Everyday Life* (Cambridge, MA: MIT Press, 2011).
18. Gregory Mougin and Elliott Sober offer a critique in "Betting against Pascal's Wager," *Noûs* 28, no. 3 (1994): 382–95.
19. There are four main reasons why the climate-change version of Pascal's Wager is more complex. First, the original wager dichotomizes both the possible states of the world (God either exists or doesn't) and the bettor's options for action (one can either wager that God exists or doesn't). In the climate-change wager, though, both the bettor's options (acting or not to address climate change) and the future state of the world (the degree of future warming and the amount of damage it causes) can't be easily dichotomized, because they're continuous variables. Second, Pascal seems to assume that heaven brings infinite pleasure (infinite positive utility, in economists' terms), while hell brings infinite displeasure (or infinite negative utility), which makes the wager's logic easier to analyze. But in the climate-change version, while the positive and negative utilities of the wager's various outcomes might be enormous, they remain finite. Third, Pascal's bettor can estimate the probability of God's existence directly, based on personal knowledge or intuition. But climate bettors who aren't themselves climate scientists must

defer to the expertise of climate scientists, which means they're implicitly estimating the likelihood the scientists are right. Their wager, then, is really about whether the scientists are right and only indirectly about the degree of future warming and the damage it might cause. And fourth, in Pascal's Wager, the cost of wagering that God exists is arguably low (perhaps, for the non-believer, a degree of moral or cognitive dissonance arising from a feeling of hypocrisy) and, importantly, that wager does not itself directly affect the reality of God's existence. But in the climate analogue, wagering that the scientists are right entails action to address climate change; that action could be costly, and the action is also precisely intended to change the likelihood of climate catastrophe.

20. When estimating the costs and benefits of a possible future event, economists take into consideration both the probability the event will occur and how far into the future it will occur. All things being equal, people weigh less heavily the potential costs and benefits of less probable events. They do the same with events expected to occur further in the future. For instance, most residents of Phoenix, Arizona, won't care much if told their region faces a devastating heat wave a century from now; but if told the heat wave will happen next month, most will care a lot.
21. In a 2019 Pew Research Center report on *Trust and Mistrust in American's Views of Scientific Experts,* 86 percent of US adults polled said "they have a great deal or fair amount of confidence in [scientists] to act in the best interests of the public," up from 84 percent in 2016. Report retrieved from https://www.pewresearch.org/science/2019/08/02/trust-and-mistrust-in-americans-views-of-scientific-experts/.
22. Paul Thagard, "Climate Change Denial: Why Don't Politicians Recognize That Humans Cause Global Warming?", *Psychology Today* (August 26, 2011), https://www.psychologytoday.com/ca/blog/hot-thought/201108/climate-change-denial.

## CHAPTER 16: HERO STORIES

1. John Dryden, "Act IV.i," *Aureng-Zebe,* ed. Frederick M. Link (Lincoln: University of Nebraska Press, 1971), 74.

2. Ernest Becker's most significant statement of his theory is *The Denial of Death* (New York: Free Press, 1973).
3. Tom Pyszczynski, Sheldon Solomon, and Jeff Greenberg, "Thirty Years of Terror Management Theory: From Genesis to Revelation," *Advances in Experimental Social Psychology* 52 (December 2015): 1–70, doi: 10.1016/bs.aesp.2015.03.001.
4. Corballis, *The Recursive Mind*, 108.
5. Becker, *Denial of Death*, 5.
6. Ajit Varki and Danny Brower, *Denial: Self-Deception, False Beliefs, and the Origins of the Human Mind* (New York: Hachette, 2013).

## CHAPTER 17: STRATEGIC INTELLIGENCE

1. All quotations of Stephanie May in this chapter are extracted from chapter 17 (starting at page 315) through chapter 22 (ending at page 418) of her unpublished memoirs.
2. Meadows, *Leverage Points*, 19.
3. Erving Goffman, *Strategic Interaction* (Philadelphia: University of Pennsylvania Press, 1970), 100-1.

## CHAPTER 18: MINDSCAPE

1. A comprehensive survey of current scholarship on political ideology is Michael Freeden, Lyman Tower Sargent, and Marc Stears, eds., *The Oxford Handbook of Political Ideologies* (Oxford: Oxford University Press, 2013). See also Teun A. van Dijk, *Ideology: A Multidisciplinary Approach* (London: Sage, 1998).
2. Foundational studies of political attitudes and ideology include Milton Rokeach, *The Nature of Human Values* (New York: Free Press, 1973); Michael Thompson, Richard Ellis, and Aaron Wildavsky, *Cultural Theory* (Boulder, CO: Westview, 1990); Valerie Braithwaite, "The Value Balance Model of Political Evaluations," *British Journal of Psychology* 89 (1998): 223–47, doi: 10.1111/j.2044-8295.1998.tb02682.x; and John T. Jost, Aaron C. Kay, and Hulda Thorisdottir, eds., *Social and Psychological Bases of Ideology and System Justification* (Oxford: Oxford University Press, 2009). In cross-national studies of political attitudes and values, of particular importance is the World Values Survey. Since 1981, the survey has gathered

large quantities of data in almost one hundred countries and is now the "the largest non-commercial, cross-national, time series investigation of human beliefs and values ever executed." For more information, see http://www.worldvaluessurvey.org/wvs.jsp. See also the "cultural dimensions theory" of Geert Hofstede, summarized in Hofstede, "Dimensionalizing Cultures: The Hofstede Model in Context," *Online Readings in Psychology and Culture* 2, no. 1 (2011), doi: 10.9707/2307-0919.1014

3. John T. Jost, Jack Glaser, Arie W. Kruglanski, and Frank J. Sulloway, "Political Conservativism as Motivated Social Cognition," *Psychological Bulletin* 129, no. 3 (2003): 339–75, doi: 10.1037/0033-2909.129.3.339.
4. An early, seminal formulation of the relevance of "moral foundations theory" to people's attachment to political ideology is Jonathan Haidt, Jesse Graham, and Craig Joseph, "Above and Below Left-Right: Ideological Narratives and Moral Foundations," *Psychological Inquiry* 20 (2009): 110–19, doi: 10.1080/10478400903028573. Haidt's most complete statement of the theory is *The Righteous Mind: Why Good People Are Divided by Politics and Religion* (New York: Pantheon, 2012). A quickly accessible treatment can be found at https://moralfoundations.org/.
5. Basins of attraction are places of local stability or equilibrium where the system in question is more likely to settle and remain, because less energy is needed to keep the system there. In the state-space model, the basins are locations in the state space where large numbers of people don't have to invest a lot of cognitive energy—they don't have to think much—to maintain their political ideologies, because (among other reasons) the ideologies at those locations align with their temperaments and moral intuitions. Psychologists can measure people's investment of cognitive energy by using methods such as the implicit association test, which captures the degree of subconscious association between mental representations in a person's mind. Generally, the more conscious an association, the more cognitive energy is invested in making the association.
6. Kevin B. Smith, Douglas R. Oxley, Matthew V. Hibbing, John R. Alford, and John R. Hibbing, "Linking Genetics and Political

Attitudes: Reconceptualizing Political Ideology," *Political Psychology* 32, no. 3 (June 2011): 369–97, doi: 10.1111/j.1467-9221.2010.00821.x.

7. Field tests so far have produced notable results. For example, in a survey of over 450 people in North America drawn from a larger group of over 5,000 to represent a cross-section of adults with different incomes, education, and liberal or conservative political orientations, the state-space questions predicted people's political orientation as well as tried-and-true approaches that social psychologists and political pollsters use—such as measures of authoritarianism, traditional conservatism, and social dominance orientation. Specifically, the questions predicted about 47 percent of the variance in political orientation in the sample (i.e., they generated together an R-squared of 0.4714 in a regression analysis). The survey also produced some surprises. For instance, while most people in the group were moderately inclined to defer to authority, most were also morally opposed to the use of power over others.
8. Jeremy Lent, "As Society Unravels, the Future Is Up for Grabs," *Patterns of Meaning*, September 12, 2019, https://patternsofmeaning.com/author/jeremylent/.
9. Rachel Carson, speaking in "The Silent Spring of Rachel Carson," CBS Reports, television documentary (April 3, 1963), as quoted in Jonathan Norton Leonard, "Rachel Carson Dies of Cancer; 'Silent Spring' Author was 56," *New York Times*, April 15, 1964, https://archive.nytimes.com/www.nytimes.com/books/97/10/05/reviews/carson-obit.html.
10. Robert Boyd and Peter J. Richerson, *Culture and the Evolutionary Process* (Chicago: University of Chicago Press, 1985); and Peter J. Richerson and Robert Boyd, *Not by Genes Alone: How Culture Transformed Human Evolution* (Chicago: University of Chicago Press, 2005).
11. Karl Jaspers, "I. The Axial Period," in *The Origin and Goal of History* (London: Routledge & Kegan Paul, 1953), 1–21.
12. Robert Bellah, *Religion in Human Evolution: From the Paleolithic to the Axial Age* (Cambridge, MA: Harvard University Press, 2011); and Karen Armstrong, *The Great Transformation: The Beginning of Our Religious Traditions* (New York: Knopf, 2006).

## CHAPTER 19: HOT THOUGHT

1. Scholars such as neuroscientist António Damásio trace this reason-emotion dichotomy back to the writings of French philosopher René Descartes; it's now so deeply embedded in modern culture that I've represented it in one of my state-space questions. In reality, however, all human cognition, including rational cognition, is "embodied," in the sense that it's intimately affected by—and in many cases only possible because of—the manifold physical properties of our bodies, including how our bodies move in and manipulate their surroundings, their sensory apparatus, and the endocrine and neurochemical systems that involve emotion. See António Damásio, *Descartes' Error: Emotion, Reason, and the Human Brain* (New York: Penguin, 2005).
2. Paul Thagard, *Brain-Mind: From Neurons to Consciousness and Creativity* (Oxford: Oxford University Press, 2019), 155. See also Thagard, *Hot Thought: Mechanisms and Applications of Emotional Cognition* (Cambridge, MA: MIT Press, 2006).
3. Thomas Homer-Dixon, Manjana Milkoreit, Steven Mock, Tobias Schröder, and Paul Thagard, "The Conceptual Structure of Social Disputes: Cognitive-Affective Maps as a Tool for Conflict Analysis and Resolution," *SAGE Open* (January–March 2014): 1–20.
4. "Some people have asked us what group we represent," Stephanie told a reporter in June 1957, when her first anti-testing petition was circulating. "There is no organization involved," she said. "The only answer is that we represent children." Quoted in "Anti-Bomb Test Petitions Circulating in Churches," *Hartford Courant*, June 8, 1957.

## CHAPTER 20: RENEWING THE FUTURE

1. May, unpublished memoirs, 446.
2. This phrasing originates with the leading US climate activist Joe Romm.
3. C.S. Lewis quoted in John Lukacs, *Five Days in London: May 1940* (New Haven, CT: Yale University Press, 1999), 199.
4. The "Moral Principles" question in the state-space list highlights the distinction between these two positions.

5. The closest analogue to my three-part distinction is the categorization system developed by the British marketing firm Cultural Dynamics Strategy and Marketing in which people are sorted by their values into groups called Settlers, Prospectors, and Pioneers. This system is itself closely related to the Theory of Basic Human Values developed by Israeli social psychologist Shalom Schwartz. See Chris Rose, *What Makes People Tick: The Three Hidden Worlds of Settlers, Prospectors, and Pioneers* (Leicester, UK: Matador, 2011); Shalom Schwartz, "Are There Universal Aspects in the Structure and Contents of Human Values?", *Journal of Social Issues* 50, no. 4 (1994): 19–45; and Shalom H. Schwartz and Anat Bardi, "Value Hierarchies across Cultures: Taking a Similarities Perspective," *Journal of Cross-Cultural Psychology* 32, no. 3 (May 2001): 268–90.
6. A key variation of this worldview—one resembling that of a violent religious cult—would have strong commitments to spirituality and to objective and universal moral principles on the Spirituality and Moral Principles dimensions respectively.
7. Amartya Sen, *Development as Freedom* (Oxford: Oxford University Press. 1999), 74–76.
8. The Stanford psychologist Albert Bandura develops a comprehensive theory of agency in "Toward a Psychology of Human Agency," *Perspectives in Psychological Science* 1, no. 2 (2006): 164–80, doi: 10.1111/j.1745-6916.2006.00011.x.
9. Herbert Kelman, "Violence without Moral Restraint: Reflections on the Dehumanization of Victims and Victimizers," *Journal of Social Issues* 29, no. 4 (1973): 25–61; and Simon Baron-Cohen, *The Science of Evil: On Empathy and the Origins of Human Cruelty* (New York: Basic Books, 2011).
10. With respect to the global crises considered in this book, I believe a deontological ethic is generally preferable to a consequentialist ethic that judges the morality of actions by the value of the outcomes or consequences those action produce for individuals and societies. Michael J. Sandel provides a brilliant overview of contending perspectives on ethics and justice in *Justice: What's the Right Thing To Do?* (New York: Farrar, Straus and Giroux, 2009).

11. Encyclical Letter *Laudato Si' of the Holy Father Francis, on Care for Our Common Home*, http://www.vatican.va/content/francesco/en/encyclicals/documents/papa-francesco_20150524_enciclica-laudato-si.html; and *The Earth Charter*, https://earthcharter.org/read-the-earth-charter/.
12. Peter Singer develops a powerful argument for global compact based on fairness between rich and poor in *One World: The Ethics of Globalization* (New Haven, CT: Yale University Press, 2002).
13. The classic study of the effect of social capital on well-being is Robert Putnam (with Robert Leonardi and Raffaella Y. Nonetti), *Making Democracy Work: Civic Traditions in Modern Italy* (Princeton, NJ: Princeton University Press, 1993). The sociologist James Coleman pioneered the development of the concept of social capital; see James S. Coleman, "Social Capital in the Creation of Human Capital," *American Journal of Sociology* 94 (1988): S95–120, https://www.jstor.org/stable/2780243.
14. Benedict Anderson, *Imagined Communities: Reflections on the Origin and Spread of Nationalism*, rev. ed. (London: Verso, 1991).
15. The challenge resembles the worldview migration problem—the difficulty of moving as a group from one worldview to another—that I described in chapter 18. Technically, however, worldview migration is a "coordination problem," much like the problem of agreeing, in the absence of government, to drive on the same side of the road. Climate change and many other global crises humanity faces, in contrast, are "collective action problems"—or, in the language of game theory, examples of the "prisoner's dilemma"—because a single actor can benefit from free-riding on others' cooperation. See Duncan Snidal, "Coordination versus Prisoners' Dilemma: Implications for International Cooperation and Regimes," *American Political Science Review* 79, no. 4 (December 1985): 923–42, doi: 10.2307/1956241.
16. Three important contributions to understanding possibilities for global cooperation among self-interested actors are Elinor Ostrom, *Governing the Commons: The Evolution of Institutions for Collective Action* (Cambridge, UK: Cambridge University Press, 1990); Scott Barrett, *Why Cooperate: The Incentive to Supply Global Public Goods* (Oxford: Oxford University Press, 2007); and Todd Sandler,

*Global Collective Action* (Cambridge, UK: Cambridge University Press, 2004).

17. Haidt, *Righteous Mind*, 260.
18. Reuters, "Young Climate Activists Accuse World Leaders of Violating Child Rights through Inaction," *World News*, September 23, 2019.
19. Samuel Johnson quoted in James Boswell, *Boswell's Life of Johnson*, ed. Charles Grosvenor Osgood, entry for September 19, 1777, https://www.gutenberg.org/files/1564/1564-h/1564-h.htm.
20. Jaspers quoted in Siep Stuurman, "1. Visions of a Common Humanity," in *The Invention of Humanity: Equality and Cultural Difference in World History* (Cambridge, MA: Harvard University Press, 2017).
21. John Gray, "Humanity Is a Figment of the Imagination. Review of Kenan Malik, *The Quest for a Moral Compass: A Global History of Ethics*," June 12, 2014, https://www.newstatesman.com/culture/2014/06/john-gray-humanity-figment-imaginatio.
22. Tom McCarthy, "Does Theresa May Really Know What Citizenship Means?", *The Guardian*, January 21, 2017, https://www.theguardian.com/books/2017/jan/21/theresa-may-citizenship-tom-mccarthy-aeschylus.
23. Émile Durkheim, the great French sociologist, usefully distinguished between mechanical and organic solidarity. Mechanical solidarity is social cohesion that arises most easily in smaller, homogenous communities without a lot of internal differentiation, where shared values and beliefs encourage community members to cooperate. Organic solidarity is social cohesion that arises in larger communities with a complex division of labor, because people's economic dependence on each other requires that they collaborate for mutual benefit. The common human "we" identity I argue for here has aspects of both types of solidarity: mechanical in its appeal to people's shared values regarding their children, and organic in its appeal to common human vulnerabilities and interdependencies around superordinate goals.
24. The Iranian American social psychologist Fathali Moghaddam uses the term "omniculturalism" (in contrast with "multiculturalism") to emphasize the principle of putting common humanity first. See Moghaddam, "Why Omniculturalism, Not Multiculturalism,

Is the Solution," *Psychology Today*, July 19, 2014, https://www.psychologytoday.com/us/blog/the-psychology-dictatorship/201407/why-omniculturalism-not-multiculturalism-is-the-solution.

25. Fritjof Capra, *The Web of Life: A New Scientific Understanding of Living Systems* (New York: Anchor, 1996), 296.
26. Daniel Wildcat quoted in Dan Zak, "How Should We Talk About What's Happening to Our Planet?", *Washington Post*, August 27, 2019, https://www.washingtonpost.com/lifestyle/style/how-should-we-talk-about-whats-happening-to-our-planet/2019/08/26/d28c4bcc-b213-11e9-8f6c-7828e68cb15f_story.html.
27. There are many commonalities between my argument here and that of the Princeton philosopher Peter Singer. Drawing on the thinking of nineteenth-century Irish historian and political theorist William Edward Hartpole Lecky, Singer argues that the ambit of human ethical concern has expanded through history, from the family and community to the nation and, now, to humanity as a whole. This expansion has happened, Singer contends, because human reason has encouraged people over time to generalize the application of biologically derived moral principles beyond their families and tribes to progressively larger groups. While I agree that the circle of our moral concern needs to be radically enlarged, my argument differs from Singer's in two key respects. First, Singer grounds his thesis in consequentialism, while mine is fundamentally deontological; I'm convinced that today's prevailing logic of economic consequentialism is a significant source of the crises humanity faces. Second, the logic of Singer's argument suggests that a common human identity—a species-wide "we"—is mostly a *result* of the expansion of moral concern, whereas I argue that this shared identity must be a *precursor to* that expansion of moral concern. We recognize our responsibility to others because we see them as part of our identity community—our "we." Put simply, I believe an extended identity creates an extended sense of justice, rather than the other way around. For the core of Singer's argument, see chapter 4, "Reason," in *The Expanding Circle: Ethics, Evolution, and Moral Progress* (Princeton, NJ: Princeton University Press, 2011 [1981]), 87–124. See also Singer's essay, "The Drowning Child and the Expanding

Circle," *New Internationalist*, April 5, 1997, https://newint.org/features/1997/04/05/peter-singer-drowning-child-new-internationalist.

28. Carl Safina, "The Real Case for Saving Species: We Don't Need Them, But They Need Us," *Yale Environment 360*, October 21, 2019, https://e360.yale.edu/features/the-real-case-for-saving-species-we-dont-need-them-but-they-need-us.
29. Stuart Kauffman, correspondence with author, November 6, 2019.
30. In the list of state-space questions, the *opportunity* commitment implies a moderate (M) or strong (S) belief strength at the "Choice" end of the Agency dimension and an M or S at the "Resist" end of the Authority dimension. The *safety* commitment implies an S at the "Dangerous" end of the Threat dimension; an M or S at the "Generous" end of the Human Nature dimension; and an M at the "Often Right" end of the Power dimension. The *justice* commitments imply an M or S at the "Objective, Universal" end of the Moral Principles dimension; an M or S at the "A Lot" end of the Care for Others dimension; and an M or S at the "Immoral" end of the Wealth dimension. And finally, the three *identity* commitments imply, respectively, an S at the "Small and Unimportant" end of the Social Differentiation dimension; an M or S at the "From One's Group" end of the Source of Personal Identity dimension; and an S at the "As One with Nature" end of the Relationship between Humans and Nature dimension.
31. The "Source of Understanding" question in the Renew the Future table no longer treats reason and feeling (where "feeling" is defined loosely as emotion and/or intuition) as separate and largely mutually exclusive ways of understanding the world. Instead, reflecting recent scientific research, the table now offers the possibility that a worldview will integrate reason and feeling as sources of understanding, which, with an S at the "Together" end of the question's scale, the Renew the Future worldview then does. If, in contrast, a worldview sees reason and feeling as distinct sources of understanding, as most conventional worldviews today still do, then a binding question must ask whether this worldview regards reason or feeling as dominant.
32. This worldview's answers to the state-space questions are, I believe, very similar to those Stephanie May would have provided, based

on what I've learned from her memoirs and conversations with her children, as described in chapter 19. This similarity suggests, once again, that Stephanie was indeed an early archetype of the kind of activist we need today.

33. The other two dotted lines (between "Inequality" and "Opportunity" and between "Scarcity" and "Rebuilding nature") carry the same implication: the emotional valences of the linked concepts are inversely correlated.

## EPILOGUE: THE BATTLE FOR TOMORROW

1. Jean-François Mouhot, "Past Connections and Present Similarities in Slave Ownership and Fossil Fuel Usage," *Climatic Change* 105 (2011): 329–55; Klas Rönnbäck, "Slave Ownership and Fossil Fuel Usage: A Commentary," *Climatic Change* 122 (2014): 1–9; and Mouhot, "Author Response to 'Slave Ownership and Fossil Fuel Usage: A Commentary' by Klas Rönnbäck," *Climatic Change* 122 (2014): 11–13.

# Illustration credits

**PROLOGUE: OUR OWN STORY**

Page 3: Kate's flower picture; Artist: Kate Homer-Dixon.

**CHAPTER ONE: SIGNALS**

Page 11: Portrait of Stephanie May; Source: Elizabeth and Geoffrey May, by permission.

Page 13: Ivy Mike, November 1952; Source: public domain.

Page 15: Stephanie May, London rally; Source: Elizabeth and Geoffrey May, by permission.

Page 19: Southern British Columbia coastline and Puget Sound from space, August 4, 2017; Source: National Aeronautics and Space Administration (NASA), Earth Observing System Data and Information System (EOSDIS), Worldview, https://worldview.earthdata.nasa.gov.

**CHAPTER TWO: HOW ABOUT. . .**

Page 30: "Less time studying"; Artist: Dan Piraro, by permission.

Page 32: Past and projected temperature changes at Earth's surface; Artist: Jacob Buurma. Source: James Hansen, et al., "Young People's Burden: Requirement of Negative CO2 Emissions," *Earth System Dynamics* 8 (2017), Figure 3, p. 581, https://doi.org/10.5194/esd-8-577-2017.

## CHAPTER FOUR: SO OUR SOULS CAN BREATHE

Page 58: "I'll take some hope"; Artist: Stephan Pastis. Pearls before Swine © 2016 Stephan Pastis. Reprinted by permission of Andrews McMeel Syndication. All rights reserved.

Page 59: "Reassuring lie"; Artist: Clay Bennett, permission from *The Christian Science Monitor* © 2006 *Christian Science Monitor*. All rights reserved. Used under license.

## CHAPTER FIVE: THE WAY HOPE WORKS

Page 62–3: Bertrand Russell letter; Source: Elizabeth and Geoffrey May. Courtesy of the Bertrand Russell Archives, McMaster University.

Page 66: Stephanie May reading; Source: Elizabeth and Geoffrey May, by permission.

Page 74: "Controlled by complexity"; Artist: Jacob Samuel, by permission The New Yorker Collection, The Cartoon Bank.

## CHAPTER SIX: IMAGINE POSSIBILITY

Page 91: Ben's shark sub; Artist: Benjamin Homer-Dixon.

Page 97: En l'an 2000: An aerial battle; Source: Public domain, https://commons.wikimedia.org/wiki/Category:France_in_XXI_Century_(fiction).

Page 97: En l'an 2000: School room; Source: Public domain, https://commons.wikimedia.org/wiki/Category:France_in_XXI_Century_(fiction).

## CHAPTER EIGHT: THE FALSE PROMISE OF TECHNO-OPTIMISM

Page 128: Sea star; Photo credit: Thomas Homer-Dixon.

## CHAPTER NINE: THE WORLD TO COME TODAY

Page 146; Maslow's hierarchy of needs; Source: A.H. Maslow, "A Theory of Human Motivation," *Psychological Review* 50 (1943): 370–96. Graphical interpretation by Thomas Homer-Dixon.

## CHAPTER TEN: A CONTEST OF WITS

Page 169–70: Lady Churchill letter; Source: Elizabeth and Geoffrey May. Reproduced with permission of the Master and Fellows of Churchill College, Cambridge.

Page 172: WITs; Artist: Jacob Buurma.

Page 173: "Never!"; Artist: Marc Roberts, now public domain.

Page 177: "Look lady"; Artist: Herb Block. A 1957 Herblock Cartoon, © The Herb Block Foundation.

## CHAPTER THIRTEEN: A MESSAGE FROM MIDDLE-EARTH

Page 217: "Lightbulbs"; Artist: Dan Piraro, by permission.

Page 218: "We have met the enemy"; Artist: Walt Kelly. © Okefenokee Glee & Perloo, Inc., by permission.

Page 221: "Tacking the ocean"; Artist: Andrei Popov, by permission.

Page 225: Enough vs. feasible, figures 1 & 2; Artist: Jacob Buurma.

Page 228: Enough vs. feasible, figure 3; Artist: Jacob Buurma.

## CHAPTER FOURTEEN: FROM GONDOR TO WASHINGTON, DC

Page 241: "Places to intervene in a system"; Source: Donella Meadows, *Leverage Points: Places to Intervene in a System* (Hartland, VT: Sustainability Institute: 1999), http://www.donellameadows.org/wp-content/userfiles/Leverage_Points.pdf.

## CHAPTER FIFTEEN: INTO THE MIND

Page 259: Calvin and Hobbes; Artist: Bill Watterson. Calvin and Hobbes © 1990 Watterson. Reprinted with permission of Andrews McMeel Syndication. All rights reserved.

Page 266: Global warming poster; Source: The Heartland Institute, by permission.

Page 273: "Documentaries"; Artists: Steve Kelley and Jeff Parker. Dustin © 2017 Steve Kelly & Jeff Parker, Dist. by King Features Syndicate, Inc.

## CHAPTER SIXTEEN: HERO STORIES

Page 279: "Consciousness"; Artist: Scott Adams. Dilbert © 2015 Scott Adams. Used By permission of Andrews McMeel Syndication. All rights reserved.

## CHAPTER SEVENTEEN: STRATEGIC INTELLIGENCE

Page 292: Elizabeth May hunger strike; Photo credit: John Lent. By permission, AP.

## CHAPTER EIGHTEEN: MINDSCAPE

Page 300: Mother Russia; Photo credit: Kroshanosha, by permission, Istock.

Page 305: State space; Artist: Jacob Buurma.

Page 315: Gestalt shift; Artist: Jacob Buurma.

## CHAPTER NINETEEN: HOT THOUGHT

Page 322 : Cognitive-affective maps, basic elements; Artist: Jacob Buurma.

Page 326: Stephanie May CAM; Artist: Jacob Buurma.

Page 328: Patrolman CAM; Artist: Jacob Buurma.

## CHAPTER TWENTY: RENEWING THE FUTURE

Page 367: Mad Max CAM; Artist: Jacob Buurma.

Page 367: Renew the Future CAM; Artist: Jacob Buurma.

# Index

**THOMAS HOMER-DIXON** holds a University Research Chair in the Faculty of Environment at the University of Waterloo, in Ontario, Canada, and is director of the Cascade Institute at Royal Roads University in Victoria, British Columbia. Between 2009 and 2014, he was founding director of the Waterloo Institute for Complexity and Innovation. Born in Victoria, British Columbia, Thomas received his BA in political science from Carleton University and his PhD from the Massachusetts Institute of Technology in international relations, defense and arms control policy, and conflict theory. His books include *The Upside of Down: Catastrophe, Creativity, and the Renewal of Civilization*; *The Ingenuity Gap: Can We Solve the Problems of the Future?*; and *Environment, Scarcity, and Violence*. His writing has appeared in *Foreign Affairs, Foreign Policy, Scientific American*, the *New York Times*, the *Financial Times*, the *Washington Post*, and the *Globe and Mail*. His current research is focused on threats to global security in the twenty-first century, including economic instability, climate change, and energy scarcity, and on how people, organizations, and societies can better resolve their conflicts and innovate in response to complex problems.